Alaska's Glaciers

Volume 9, Number 1 / 1982 / Alaska Geographic®

The Alaska Geographic Society

To teach many more to better know and use our natural resources

ALASKA GEOGRAPHIC®, ISSN 0361-1353, is published quarterly by The Alaska Geographic Society, Anchorage, Alaska 99509-6057. Second-class postage paid in Edmonds, Washington 98020-3588. Printed in U.S.A. Registered trademark: Alaska Geographic. ISSN 0361-1353; Key title Alaska Geographic.

THE ALASKA GEOGRAPHIC SOCIETY is a nonprofit organization exploring new frontiers of knowledge across the lands of the polar rim, learning how other men and other countries live in their Norths, putting the geography book back in the classroom, exploring new methods of teaching and learning — sharing in the excitement of discovery in man's wonderful new world north of 51°16′

MEMBERS OF THE SOCIETY RECEIVE *Alaska Geographic®*, a quality magazine which devotes each quarterly issue to monographic in-depth coverage of a northern geographic region or resource-oriented subject.

MEMBERSHIP DUES in The Alaska Geographic Society are $30 per year; $34 to non-U.S. addresses. (Eighty percent of each year's dues is for a one-year subscription to *Alaska Geographic®*.) Order from The Alaska Geographic Society, Box 4-EEE, Anchorage, Alaska 99509-6057; (907) 274-0521.

MATERIAL SOUGHT: The editors of *Alaska Geographic®* seek a wide variety of informative material on the lands north of 51°16′ on geographic subjects — anything to do with resources and their uses (with heavy emphasis on quality color photography) — from Alaska, Northern Canada, Siberia, Japan — all geographic areas that have a relationship to Alaska in a physical or economic sense. In late 1981 editors were seeking material on the following geographic regions and subjects: Alaska fish and fisheries, the Seward Peninsula, and mining in Alaska. We do not want material done in excessive scientific terminology. A query to the editors is suggested. Payments are made for all material upon publication.

CHANGE OF ADDRESS: The post office does not automatically forward *Alaska Geographic®* when you move. To insure continuous service, notify us six weeks before moving. Send us your new address and zip code (and moving date), your old address and zip code, and if possible send a mailing label from a copy of *Alaska Geographic®*. Send this information to *Alaska Geographic®* Mailing Offices, 130 Second Avenue South, Edmonds, Washington 98020-3588.

MAILING LISTS: We have begun making our members' names and addresses available to carefully screened publications and companies whose products and activities might be of interest to you. If you would prefer not to receive such mailings, please so advise us, and include your mailing label (or your name and address if label is not available).

Library of Congress cataloging in publication data
Molnia, Bruce.
Alaska's glaciers.

(Alaska geographic, ISSN 0361-1353; v. 9, no. 1)
Text prepared by Bruce Molnia.
1. Glaciers—Alaska. I. Title. II. Series.
F901.A266 vol. 9, no. 1 517.98s 81-20508
[GB2425.A4] [551.3′12′09798] AACR2
ISBN 0-88240-167-X

STATEMENT OF OWNERSHIP MANAGEMENT and CIRCULATION

(Required by 39 U.S.C. 3685)

Alaska Geographic® is a quarterly publication, home offices, Box 4-EEE, Anchorage, Alaska 99509. Editor is Robert A. Henning. Publisher is The Alaska Geographic Society, Box 4-EEE, Anchorage, Alaska 99509. Owners are Robert A. Henning and Phyllis G. Henning, Box 4-EEE, Anchorage, Alaska 99509. Robert A. Henning and Phyllis Henning, husband and wife, are owners of 100 percent of all common stock outstanding.

Alaska Geographic® has a paid circulation of 18,410 subscribers and newsstand buyers.

I certify that statements above are correct and complete:
ROBERT A. HENNING
Editor

The Cover — *Margerie Glacier in Glacier Bay National Park and Preserve is very active, and rises from tidewater in a series of terraces on the flanks of Mount Fairweather, 15,300 feet elevation.* (Staff photo)

Title page — *Decapitated and insulated with a thick layer of brown volcanic ash, the Knife Creek Glaciers lie dormant near the head of the Valley of 10,000 Smokes, in Katmai National Park and Preserve. Only the rim of Mount Katmai's caldera is visible on the skyline where once stood a peak some 2,000 feet taller.* (Sean Reid, staff)

Right — *A group of skiers pauses on the lower slopes of Mount McKinley to enjoy the beauty of the alpenglow. This photo was taken from Ruth Amphitheater, at the head of Ruth Glacier.* (Third Eye Photography)

About This Issue: Bruce Molnia of the U.S. Geological Survey provided the main text and many of the photos (unless otherwise noted) for this issue. Dr. Molnia, for several years a lecturer at the Juneau Icefield Research Program, has been studying the geological forces at work in Alaska for many years and is currently project chief for environmental studies of the survey's Eastern Gulf of Alaska Project.

We are grateful to Dr. R. Dale Guthrie, a professor in the biology department of the University of Alaska, and to his wife, Mary Lee Guthrie, writer and artist who works at home, for sharing their knowledge of Ice Age Alaska.

A special thanks goes to Dr. Cyrus Peck, Sr., Cyrus Peck, Jr., and radio station KTOO-FM, Juneau, for their contribution on the Tlingit legend; and to Harvey Bowers, a geologist/hydrologist for an Anchorage consulting engineering firm, whose text sheds light on the life forms of Alaska's glaciers. We appreciate Janet Cogan and Ron Rose sharing their experiences at Child's Glacier, and finally, we thank the fine photographers whose material is included here.

Maps on pages 66, 67, 68, 82, 83, 84, 92, 93, 94, 105, 106, 107, 116, 117, 122, 123, and 130 are courtesy of American Geographical Society, New York, New York, and Cold Regions Research and Engineering Laboratory, Corps of Engineers, U.S. Army, Hanover, New Hampshire.

Editors: Robert A. Henning, Barbara Olds, Penny Rennick
Editorial Assistant: Kathy Doogan
Designer: Sandra Harner
Cartographer: Jon.Hersh

Editor's note: *Items in* **boldface** *in the text refer to entries in the glossary where further information about that particular word will be found. The item is boldfaced where it is discussed in the text, not necessarily where it first occurs.*

Above — *A backpacker approaches the face of Exit Glacier, near Seward. Many of Alaska's glaciers are easily accessible to those who wish to get a good, close-up look.* (Bill Kossen, *Anchorage Times*)
Right — *Two skiers follow the trail of the rest of their group along a ridge, approaching a pass from which they will drop onto Ruth Glacier, in Denali National Park and Preserve.* (Mark Skok)

Contents

Rivers of Ice

Alaska is a land of many contrasts: majestic mountains, endless miles of shoreline, and thousands of glaciers. Alaska is not unique in the United States, as more than 1,000 glaciers, covering an area of about 205 square miles, exist in Washington, Wyoming, Montana, Oregon, California, Colorado, Idaho, and Nevada. All of the glaciers in these states combined, however, could not even equal the area of a large Alaska valley glacier system. Alaska's glaciers range from tiny cirque glacier remnants covering fractions of a square mile, to massive piedmont glaciers each covering hundreds of square miles.

Tourists rarely get to see the largest of Alaska's glaciers, except perhaps from the window of an airplane going to or from Anchorage, Juneau, or Fairbanks. Rather, the majority of visitors see a variety of medium- to small-sized valley glaciers — Mendenhall, Matanuska, Portage, or Worthington — that are accessible by car or bus, or they see the glaciers of Glacier Bay or Columbia Glacier from the deck of a ship.

The few who are fortunate enough to explore the surface of a glacier, or to linger in front of a calving tidewater

Left — *Mendenhall Glacier, located about 13 miles from downtown Juneau, is the most visited glacier in Alaska. Mendenhall Lake, at the glacier's terminus, is constantly enlarging as the ice retreats.*
Right — *Many of Alaska's smaller glaciers, such as these two unnamed tributaries which join to form Grand Plateau Glacier in the Fairweather Range, can be seen from the air. Air charter services provide opportunities to fly over many glacier-covered areas of the state.*

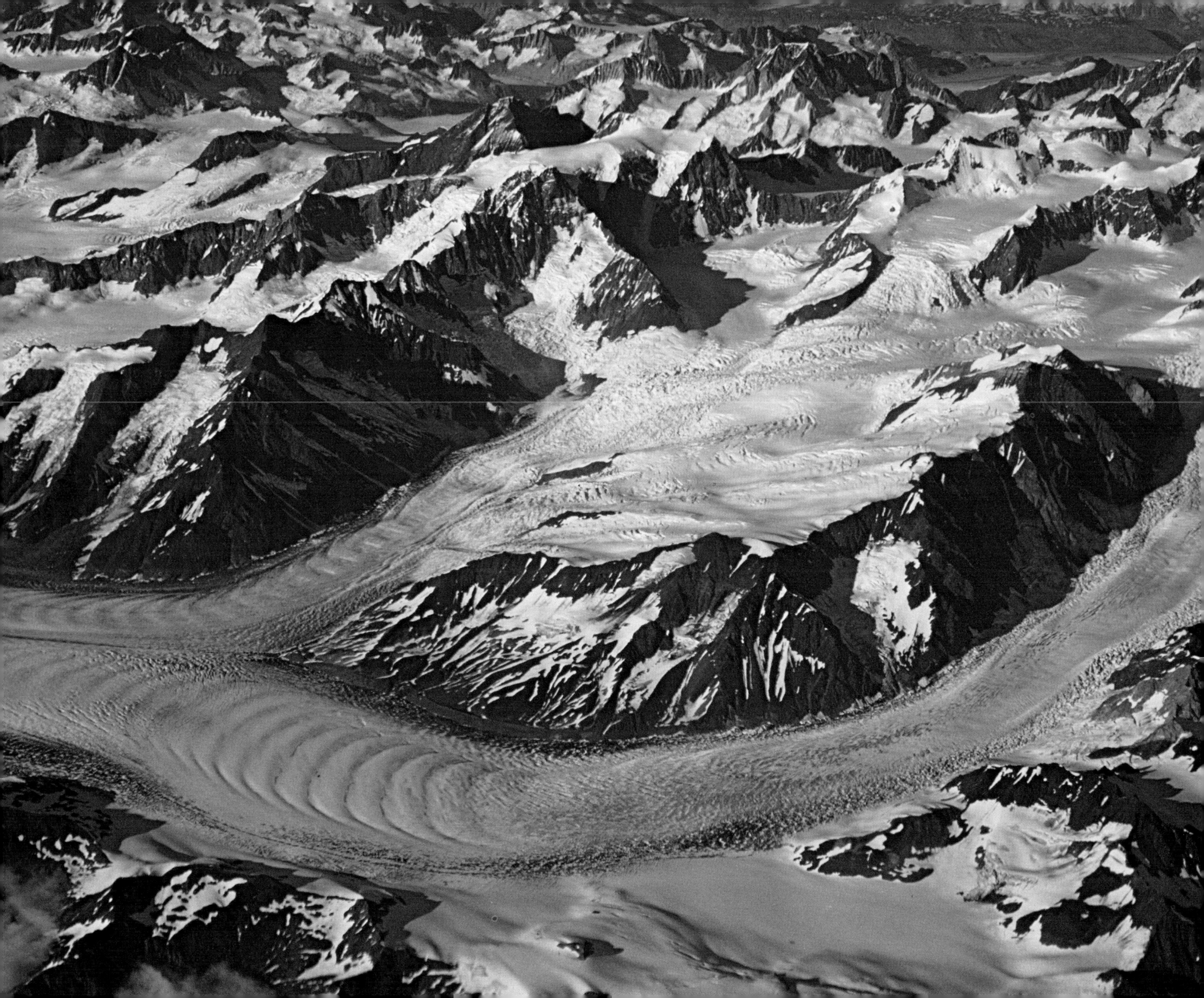

glacier, take home memories that last a lifetime. This book presents the story of Alaska's glaciers to those many who have not had the opportunity to examine and explore firsthand the majesty and power of a living Alaska glacier. To those few who have a favorite glacier or ice field, I want to apologize if I do not give it the coverage and attention you feel it deserves. Only so many glaciers can be described within the limits of this volume, and in a land where the number of glaciers is in the tens to hundreds of thousands, the overwhelming majority must unfortunately be omitted. I also have my favorites and as I describe them, I will try to present general background and basic geological information that can be applied to all of Alaska's glaciers.

We are lucky in that we live in a period of time, both historical and geological, when glaciers exist and when we have the technical ability to travel to where they are, to photograph them, and to share this knowledge with others. Less than 500 years ago much of the scientific world was ignorant of the existence of glaciers. Less than 200 years ago the majority of the geological community failed to recognize the importance of glacial erosion, and slightly more than a century ago more than 99% of Alaska's glaciers were completely unknown. Now, however, many of the questions about why glaciers form, how they move, and what they do are being answered. I hope this book will explain many of these phenomena to you.

Bruce Molnia

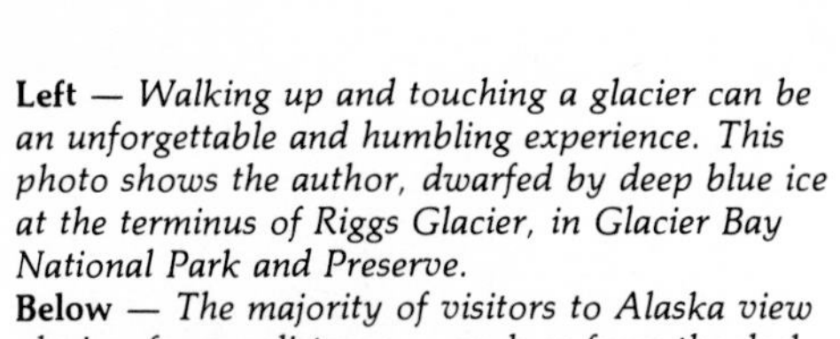

Left — *Walking up and touching a glacier can be an unforgettable and humbling experience. This photo shows the author, dwarfed by deep blue ice at the terminus of Riggs Glacier, in Glacier Bay National Park and Preserve.*
Below — *The majority of visitors to Alaska view glaciers from a distance — such as from the deck of a cruise ship. Here, the* Island Princess *carefully makes its way through waves and icebergs to give its 500 passengers a closer look at some of the glaciers in Glacier Bay National Park and Preserve.*

Above — *Some of Alaska's glaciers are easily accessible by car, such as Worthington Glacier, which can be seen along the Richardson Highway, about 20 miles north of Valdez. Huge, fresh lateral and terminal moraines surround the glacier, indicating that it is retreating.* (John Koranda)

Anatomy of a Glacier

A **glacier** is a perennial accumulation of ice, snow, water, rock, and sediment that moves under the influence of gravity. A complex and dynamic system, glaciers continuously change in response to fluctuations in temperature, precipitation, and other geologic processes.

To understand the working of a glacier, we must carefully define the component parts of the glacier system. These are threefold: first is the glacier ice and the material entrained in and on it; second is the glacial valley, fiord, or channel, and its related rock features that the ice mass flows in, on, and over; third is the complex array of deposits that are produced by the glacier as it advances, retreats, or melts in place.

Left — *Following the long process through which snow is changed to glacier ice, individual snowflakes merge to form granules of solid ice. Some ice crystals may grow to be longer than one foot.*
Below — *The point where a glacier ends, whether it is on land or water, is called the terminus. The most spectacular termini are those which calve icebergs into the water, such as Muir Glacier, in Glacier Bay National Park and Preserve.*

Glacier Ice

Glacier ice forms through a slow continuous change of snow to a material called **firn** and finally to bubbly glacier ice. This change takes place as successive years' snowfalls pile up as layers on each other, increasing the pressure on the older layers as they are buried. This causes changes in density, volume, and crystal structure. New fallen snow has a density of 0.88 to 0.92. (This also explains why glacier ice is blue — the physical characteristics of the water molecules absorb all colors except blue, which is transmitted.) Thus, in the metamorphic process that

changes snow to ice, a volume decrease of up to nine times may occur. The change from snow to glacier ice may take many years and may be completely stopped and destroyed by an increase in temperature or melting. Snow becomes firn after it survives one summer melt season. Firn becomes glacier ice when its permeability to air and water is zero, or when fluids cannot flow through it. As firnification (the process of changing snow to firn to glacier ice) occurs, original layering is lost and individual snowflakes merge and grow together to form granules of solid ice. Some single ice crystals may grow to more than a foot in length. Sediment deposited on a glacier's surface by running water, from avalanches, or through the air as dust falls or volcanic ash, also accumulates as distinct layers among the snow, firn, and ice. These layers frequently undergo much deformation and contortion during the firnification process. The density of firn ranges from about 0.4 to 0.8 grams per cubic cen-

This aerial view of an unnamed valley glacier in the Chugach Mountains illustrates many major glacial features. The terminus of the glacier, which is receding, is covered with debris. Five distinct medial moraines can be seen reaching toward the terminus. On the right valley wall are a well-developed lateral moraine and two trimlines. The firn line, the point of transition between bare glacier ice and last year's snowfall, is clearly shown just above the center of the photo. Also easily identified are three separate tributary glaciers, several cirques, and an arête ridge.

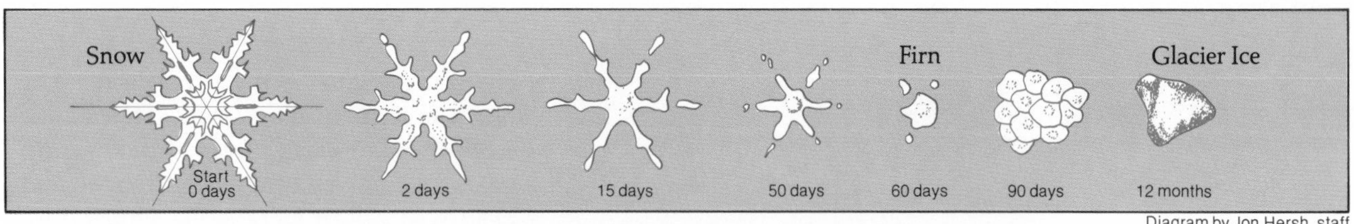

Figure 1 — *Stages in the transformation of a snowflake to granular ice (firn) to glacier ice. Source: modified from U.S. Army Corps of Engineers*

timeter. At one time, the French term **névé** was used interchangeably with firn, but today *névé* is restricted to mean an area covered with perennial snow, or the area of accumulation at the upper end of a glacier.

A glacier's surface can be divided into two zones: one where more snow accumulates most years than is lost to melting, evaporation, or sublimation; and a second zone where more snow and ice is lost to these processes than accumulates. The part of the glacier with the increase in snow and ice is the accumulation area; the part showing a decrease in snow and ice is the **ablation** area or the area of wasting. The line where these two zones meet, where the annual loss is equal to the annual gain, is the equilibrium line which is often close to the firn limit, the edge of snow covering on a glacier's surface at the end of a summer season.

Glaciers move down slope under the influence of gravity. A typical valley glacier will head, or begin, on the side of a mountain at an elevation of several thou-

This massive iceberg, which rises more than 30 feet out of the water, shows well-defined layers of accumulated sediment.

sand feet or more. The glacier will flow down the side of the mountain to a much lower elevation where its end, or **terminus,** will be located. Many Alaska valley glaciers reach sea level and have their termini discharging into bays or fiords of the Pacific Ocean.

If the distance between the head of the glacier and its terminus is increasing, we describe the glacier as advancing. If the distance is decreasing, the glacier is said to be retreating. Regardless of whether a glacier is retreating or advancing, the ice within the glacier is continuously moving down slope. A delicate balance exists between the rate that ice is moving down the glacier and the rate that melting or calving (with glaciers that end in bodies of water) is taking place. If the quantity of ice lost to melting and calving is greater than the amount of ice coming from above, then the glacier will retreat; when the supply is greater than the loss, then the terminus will advance. When the supply is equal to the loss, then the position of the terminus will remain stationary. Even when a glacier's terminus is rapidly retreating, the ice in that glacier is still moving downhill toward the terminus.

Surging Glaciers

Discussion of advancing and retreating glaciers deals specifically with changes in the position of a glacier's terminus. A separate, but very important additional type of motion known as **surge** is used to describe sudden, large-scale, short-lived movements within a glacier, regardless of whether a change in the position of the glacier's terminus occurs. Surges are periodic and generally involve ice movements 10 to 100 times faster than the glacier's normal flow rate. Austin Post of the U.S. Geological Survey attributes these rapid movements to a "remarkable instability which occurs at periodic intervals in certain glaciers." Other investigators have suggested that certain surges were the result of earthquakes, avalanches, or local increases in snow accumulation.

Based on aerial photographic observations, Post has identified 204 surging glaciers in North America, at least two-thirds of which are in Alaska. These glaciers were identified on the basis of intense crevassing and rapid ice displacements, characteristics of surges; or distinctive surface features that resulted from previous surges. Post found that surging glaciers occur in specific areas. In Alaska they are restricted to the Alaska Range, eastern Wrangell Mountains, eastern Chugach Mountains, and the Saint Elias

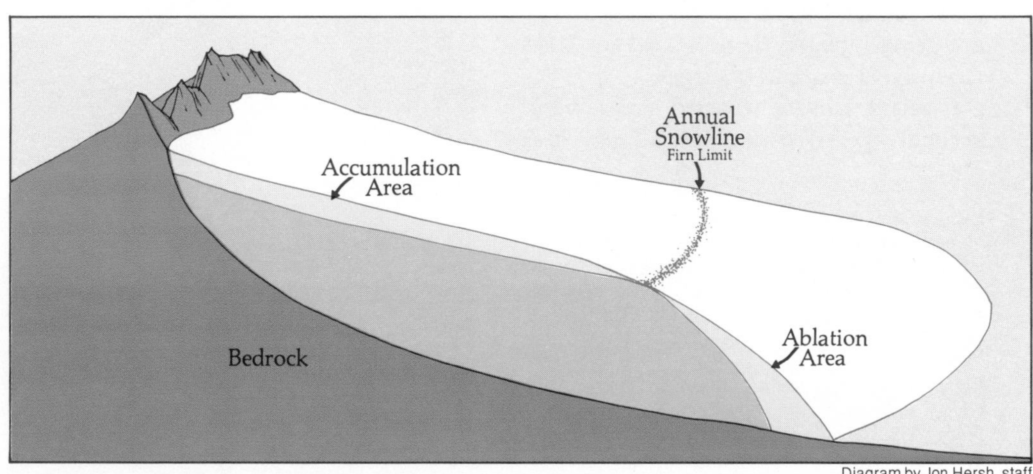

Diagram by Jon.Hersh, staff

Figure 2 — *This theoretical cross section of a glacier shows the relationship of accumulation and ablation areas and the annual snow line (firn limit). Also shown is a plan view of a glacier with the ablation and accumulation areas outlined.*

One method of determining the health of a glacier is by calculating the Accumulation Area Ratio or AAR. The AAR is defined as the ratio of the accumulation area to the total area of the glacier. The closer the AAR number is to one, the healthier the glacier.

Mountains near Yakutat and Glacier Bay. No surging glaciers have been identified in the Coast Mountains, west and central Wrangells, west and central Chugach Mountains, Kenai Mountains, or the Brooks Range. Post's analysis of surging glaciers showed that they exist in maritime to continental climates and in temperate to subpolar environments. Surging is independent of elevation, bedrock type, valley configuration, glacier orientation, or size. The two largest glaciers in North America, Malaspina and Bering, both show evidence of multiple surges.

Post suggests that possible causes of surges may be related to bedrock roughness or permeability, uncommonly high ground-water temperature and abnormal geothermal heat flow. During its 1956 to 1957 surge, Muldrow Glacier moved as much as 1,150 feet per day. Today, Variegated Glacier in Russell Fiord and Black Rapids Glacier in the Alaska Range are being intensely investigated for clues to unravel the origin of surges.

Glaciers can be likened to a bank account. When over a period of years more snow and ice accumulates in the zone of accumulation than is lost to melting, we refer to a glacier as being "healthy" with a positive budget. Such glaciers are generally advancing. Glaciers with a negative ice budget are generally retreating, and are euphemistically referred to as being "sick."

Alaska glaciers vary in thickness from a few hundred feet to more than 2,000 feet. Flow within a glacier varies according to depth and thickness of ice, shape of the channel or valley, and temperature. As a glacier flows, the upper 100 to 150 feet

Left — *Deep cracks, known as crevasses, develop along the surface of the glacier, some reaching lengths of thousands of feet. Inside these crevasses, experienced climbers find an eerie silence, broken only by an occasional gurgle of meltwater or moaning sound as the walls shift.*
(Ransom Saltmarch)
Right — *The upper surface of Johns Hopkins Glacier is covered with jagged parallel crevasses. Since 1935, the glacier has advanced more than two miles.*

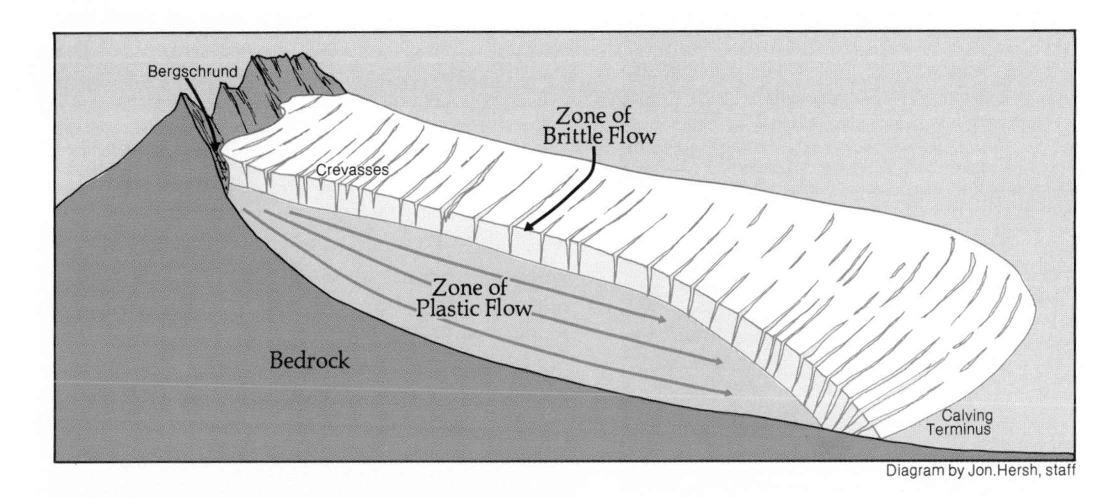

Figure 3 — *Zones of brittle and plastic flow, and longitudinal lines of flow, are shown in this diagram of flow patterns in a glacier. Note that crevasses do not extend below the zone of brittle flow.*

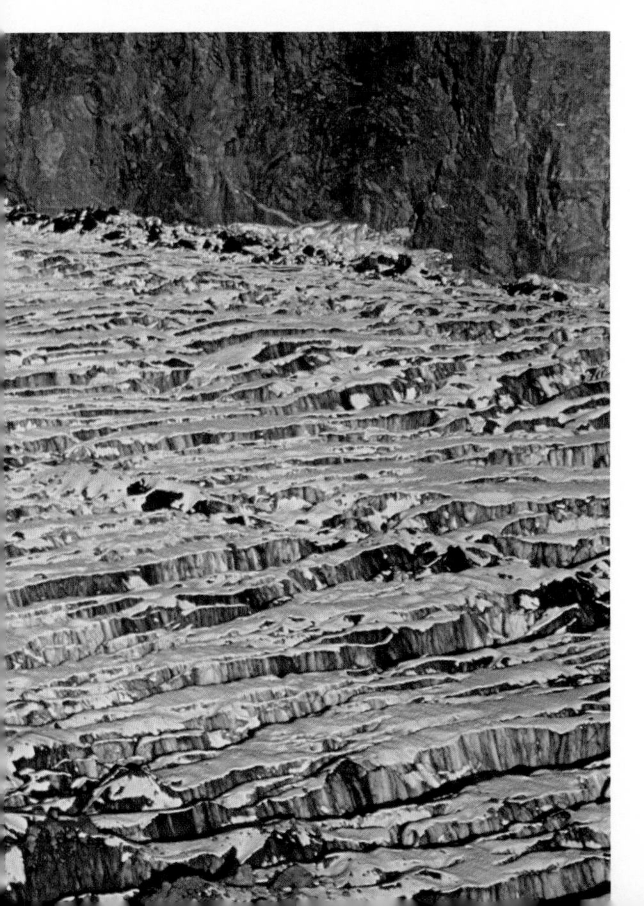

deforms and flows in a brittle fashion, often developing elongate cracks that may extend for hundreds to thousands of feet across the glacier's surface. The cracks, which often change in size and shape as the glacier flows downvalley, are called **crevasses.** Below a depth of 100 to 150 feet, the glacier is under much more pressure than at the surface and behaves more plastically.

Many glacier travelers have lost their lives from falling into crevasses, which often are hidden under snow. At the very head of a valley glacier a single large crevasse or series of smaller crevasses develops at the location where moving ice pulls away from the rock or cirque wall against which it has accumulated. This crevasse system is called a **bergschrund.**

Because of differences in glacier motion and flow, ice crystals of different size and type are segregated into bands or layers called **folia,** with the layering or banding termed **foliation.** Often, sequences of folia consist of alternating layers of clear and bubbly ice, or alternating layers of fine-grained and coarse-grained ice. Rarely does the foliation correspond to the original snow and firn accumulation layers.

Ogives are curved bands or undulations in the ice on a glacier's surface which occur in patterns generally oriented in a convex downglacier position. Ogives occur as two types: **band ogives**, alternating light and dark bands on a flat, smooth, glacier surface; and **wave ogives**, undulations of varying height in the surface of the ice. Ogives form below icefalls, areas where glaciers cascade over steep sub-ice bedrock slopes. Various theories have been presented to explain the origin of ogives. Most center around differences in rate of flow of ice over an icefall in

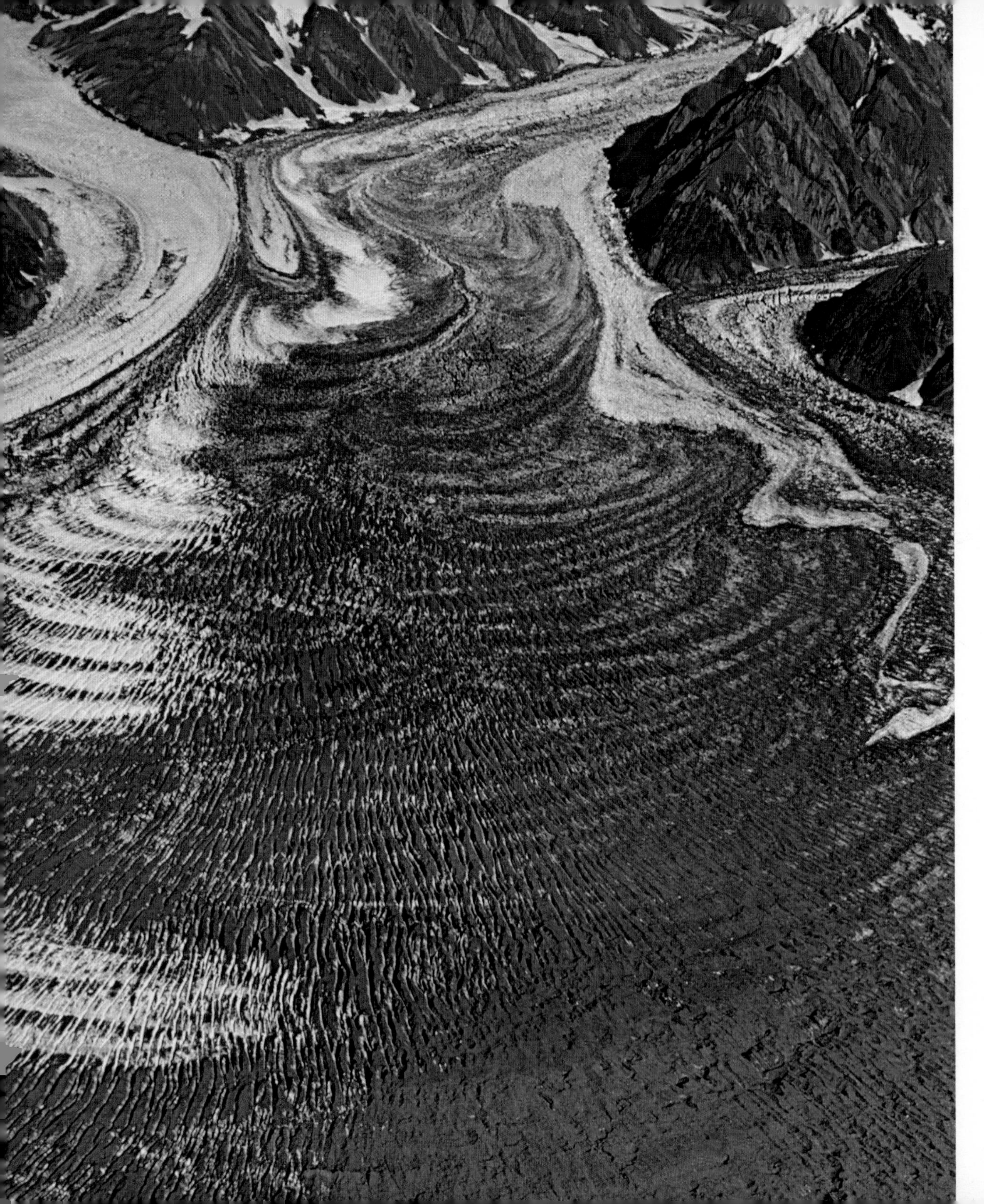

summer versus winter. One pair of dark and light bands or one pair of large and small bands represents the total flow of a full year.

Running water is an integral part of glacier ice. Streams exist in, on, and under most Alaska glaciers. During summer much of the surface of the ablation zone is wet with streams flowing into crevasses or narrow tubular chutes called **moulins** or glacier mills. Running water frequently melts channels or valleys into the surface of a glacier. Larger channels may be 15 feet wide or more.

Bergy Seltzer or Ice Sizzle

Air trapped within glacier ice is frequently subject to pressures that may exceed 750 pounds per square inch. When glacier ice is released as icebergs at near sea level pressures, many of the bubbles contained in the bergs still are in equilibrium with the depths and pressures under which they formed. As melting progresses, the pressurized air bubbles break,

Left — *Ogives, bands or wavy forms on the surface of the glacier, and moraines create spectacular patterns on Marvine Glacier. Band ogives are alternating light and dark bands on a smooth glacier surface and wave ogives are undulations of varying height in the surface of the ice.*
Right — *Icefalls occur when glaciers overflow mountaintops or ridges and cascade downhill. This icefall originates from Vaughn Lewis Glacier, in the Juneau Icefield.*

producing a resounding pop. When hundreds of these bubbles break almost simultaneously the resultant crackling sound, reminiscent of Rice Krispies cereal, is known as **ice sizzle** or **bergy seltzer.**

Intense energy is released through ice sizzle. Sound generated by ice sizzle has been detected at distances of up to 100 miles by underwater hydrophones, sound detection devices used in tracking distant ship motions. Glaciologists laughingly tell the story of the celebration of the completion of a deep ice core that was drilled to a depth of more than 2,000 feet. Ice from the bottom of the core was used at the party to chill the liquid refreshments. The celebration quickly turned into disaster as glass after glass cracked under the impact of the bursting high pressure bubbles.

To experience ice sizzle on a greatly reduced scale the only ingredients needed are a glass of water and a refrigerator ice cube. Although there is no pressure difference between the water and the bubbles in the ice cube, small audible pops can be heard as the ice cube melts. As melting continues, cracks may develop along bubble planes and the ice cube may split or fracture. In a similar fashion, glacier icebergs may fracture or roll over following intense episodes of ice sizzle.

Even icebergs stranded high and dry by changes in tides can be heard popping as

Below — *In the upper ends or sides of glacier valleys, rounded basins, called cirques, are found. This cirque, located in the Saint Elias Mountains, has a well-developed* bergschrund — *a crevasse which develops when the moving ice pulls away from the cirque wall.*
Right — *Rock flour, small rock particles scraped away by the underside of a glacier as it moves, is dumped into Johns Hopkins Inlet from Tyeen Glacier. Glaciers are thought to be nature's most efficient mechanism for erosion.*

they sit on the sand. If you ever have the opportunity to be near an iceberg, listen closely and you will hear the symphony of bubbles as they play the themes of ice sizzle.

Glacier Valley and Bedrock Features

Wherever glaciers flow, topography is changed. Glaciers tend to round, deepen, oversteepen, and streamline the surfaces they contact.

Glaciers may be the most efficient mechanism for erosion on earth. As glaciers slide over bedrock at their base, they carry and drag a load of sediment and rock known as the basal till layer. This layer, which is frozen into the basal ice, polishes and scrapes away small rock particles, known as **rock flour,** through the process of abrasion.

In addition to abrasion, which produces material fractions of an inch to inches in size, glaciers quarry or lift and carry away large blocks of bedrock that have been prepared for transport by the freezing and thawing of water in cracks, joints, and fractures. This transportation of larger blocks is known as glacial **plucking.** Plucked blocks are often cobble or boulder size but may reach a maximum dimension greater than 25 feet.

Glaciers generally flow along the path of least resistance. It is not uncommon for a glacier to occupy a stream valley or a fault trench and to widen and modify it by abrasion and plucking. Streams flow in V-shaped valleys. Glaciers modify the shape of these stream valleys and change them into a characteristic U-shaped glacial trough or valley. In the process of widening and deepening coastal valleys, glaciers often erode the valley floor to below sea level and extend the valley onto the continental shelf or into deep bays. These valleys are termed **fiords.** Alaska has hundreds of excellent examples of fiords, the best known occurring in the Inside Passage and Glacier Bay of southeastern Alaska and in Prince William Sound. Large fiords may be up to five miles in width and have more than a mile of vertical relief.

Glacier valleys are not only rounded or U-shaped in cross section, but they also have rounded, amphitheater-like basins cut in their upper ends or sides. The half-bowl-shaped depressions, called **cirques,** have lips at their lower ends, called thresholds. Small glaciers with cirques and thresholds high on the walls of large U-shaped valleys, hundreds of feet above the surface of the main valley glacier, may have tongues of ice cascading or even avalanching down the valley walls until they reach the main glacier. Such glaciers are called **hanging glaciers** and their valleys are called **hanging valleys.**

Often, valley glaciers develop in closely spaced parallel valleys. As individual

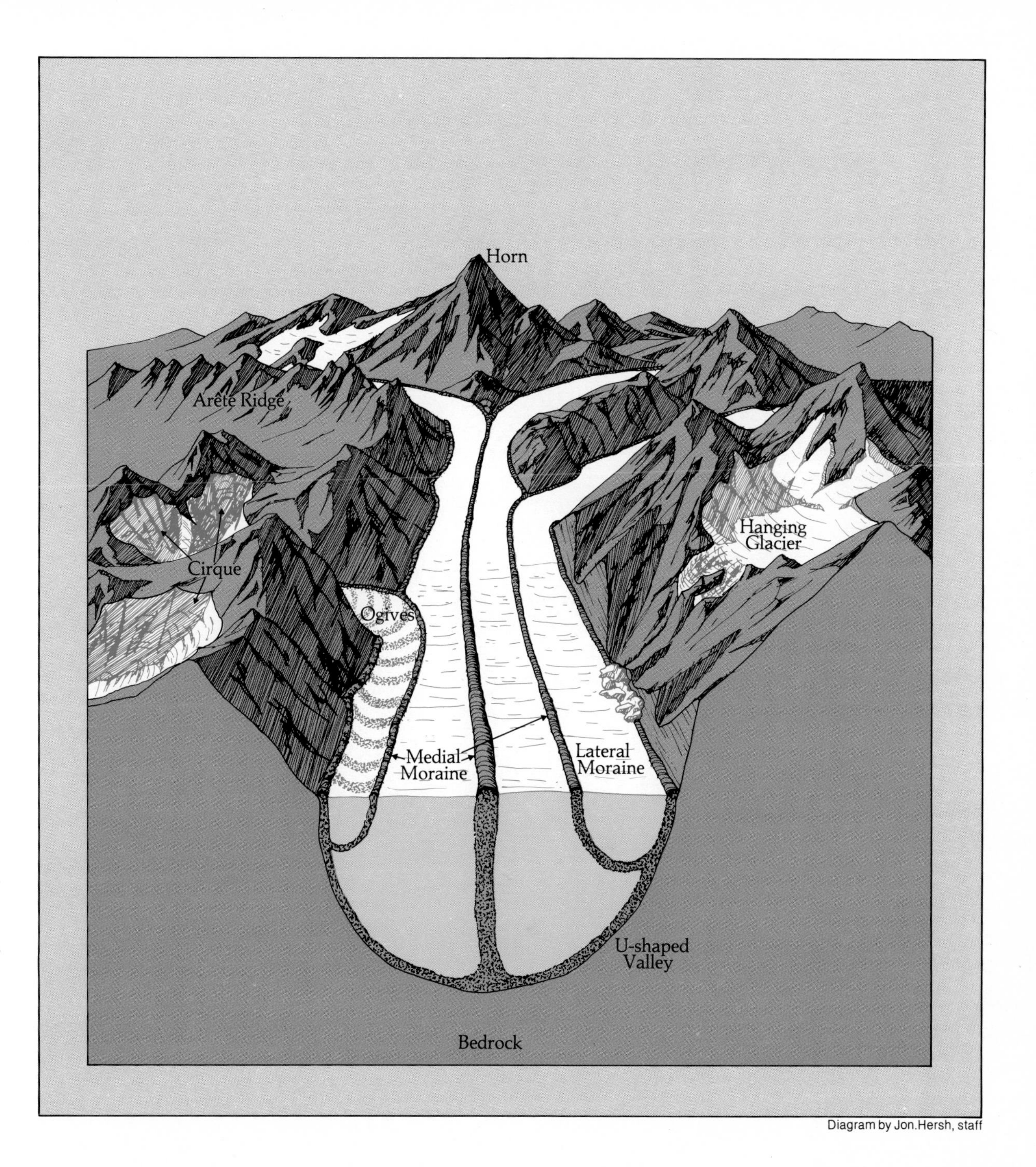

Figure 4 — *Component parts of a valley glacier system.*

Hoonah Glacier, a typical hanging glacier, cascades down a hill to reach the waters of Johns Hopkins Inlet. The glacier was named for the Hoonah tribe of Tlingit Indians.

glaciers erode and widen their valleys, the bedrock ridge that separates the two adjacent glaciers narrows and becomes oversteepened. A series of narrow, jagged spires, much like the edge of a serrated knife blade, develops along the crest of the narrowing ridge. These features are called **arêtes,** after the French word for fish bones. Eventually, with continued erosion, the *arête* ridge will disappear and the glaciers will merge.

As glaciers erode headward, or upvalley, into the side of a mountain, they modify the configuration of the mountain's summit to form a steep-sided, sharp-pointed pyramidal mountain peak, called a **horn.** When four glaciers erode a symmetrically shaped horn, it is termed a **matterhorn** after the famous peak in the Swiss Alps. Horns, *arêtes,* and cirques are the most common bedrock features that can be observed in recently deglaciated parts of Alaska, and are also often seen poking through the edges of Alaska's many valley glaciers.

As a glacially eroded area emerges from melting ice, many large, rounded, asymmetrical bedrock knobs, called **roches moutonnées** (after the French term for grazing sheep; literal translation is "fleecy rocks"), begin to emerge. These knobs, which form subglacially, have a gentle slope on their upglacier side and a steep to almost vertical face on their down-glacier side. This geometry is due to the glacier gliding and gently overriding the near side and then plucking blocks of rock from the knob's backside as it flows past. Glacier

When valley glaciers form in parallel, closely spaced valleys and then erode, widening their valleys, the bedrock ridge separating the glaciers narrows and becomes a series of jagged spires, called an arête. *This* arête *ridge is located in the Chugach Mountains, east of Anchorage.*

geologists use the shape and placement of *roches moutonnées* to help them interpret glacier flow directions.

When a glacier thins, as melting increases or as a recession begins, a sharp boundary line that marks the maximum extent of the glacier's margin appears. This line, which may either be a change in type or presence of vegetation or a change from weathered to unweathered bedrock, is called the **trimline**. Comparing the height of the most recent trimline with the surface of a glacier is very useful in determining the state of the glacier's health.

As glaciers move over their beds, they often polish, groove, or striate the surface. When ice motion is irregular, large particles may grind along the bottom and carve crescent-shaped fractures, gouges, or chatter marks into the bedrock surface, perpendicular to the direction of glacier flow. Striations and grooves are long, straight, parallel furrows, oriented in the direction of glacier movement.

Glacial Deposits

The term **drift** is applied to all sediment deposits (mud, sand, gravel, or boulders) that are transported by, or deposited directly by a glacier, or deposited by running water that originated from a glacier.

Till, sediment deposited directly by ice, is generally poorly sorted and not layered. There are two types of till: **lodgement till,** material plastered in place by ice as it moves; and ablation till, material dropped to the ground as stagnant ice melts in place. After a glacier has melted, a blanket of till, also called **ground moraine,** covers all surfaces over which the glacier flowed.

If the terminus of a glacier remains essentially in the same place for a period of time, a mound of till known as a moraine will develop around the perimeter of the glacier. If this moraine is the farthest down valley, it is called the **terminal,** or end, **moraine.** Successive moraines up a glacial valley, representing positions where the ice margin temporarily stood still during a period of retreat, are termed **recessional moraines**. Both terminal and recessional moraines often dam meltwater streams, forming moraine-dammed glacial lakes.

Jokulhlaups are outburst floods that form through the failure of glacier ice-dammed lakes. The term *jokulhlaup* is adopted from the Icelandic language. Iceland has a detailed, recorded history of periodic glacier floods that extends back to the 14th century.

Two other terms applied to types of

Right — *Kettles are formed when a retreating glacier leaves blocks of ice which become covered with sediment. As the ice melts, the sediment sinks to form depressions, such as these found on the ablation moraine of Marvine Glacier.*

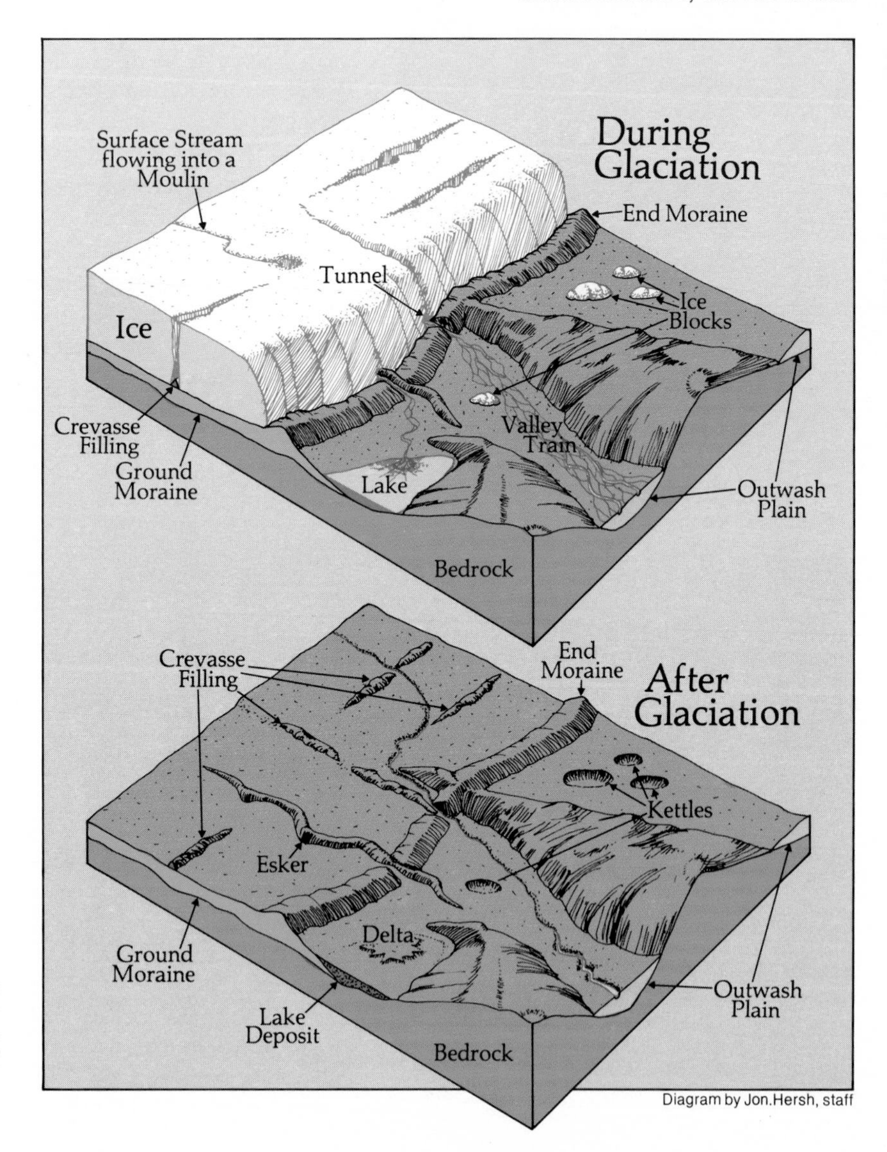

Figure 5 — *These two diagrams show the deposits of a glacier during and after glaciation. As a glacier retreats, crevasses fill and eskers rise from ground moraine.*

moraines that occur on the surface of a glacier are **lateral moraines** and **medial moraines.** A lateral moraine occurs on each side of a valley glacier and is composed of abraded sediment and plucked rock material from the side valley walls or rock and sediment that avalanche onto the ice's surface. Where two valley glaciers unite, lateral moraines merge to form a medial moraine, a moraine that is now in the middle of the combined glacier. Lateral moraines are frequently preserved after a glacier melts away, generally as a veneer of sediment plastered on the valley wall. Medial moraines are rarely preserved, as meltwater streams rework them and transport away much, if not all, of the sediment. Large valley glaciers or piedmont glaciers may show 20 or more distinct medial moraines, indicating the enormous

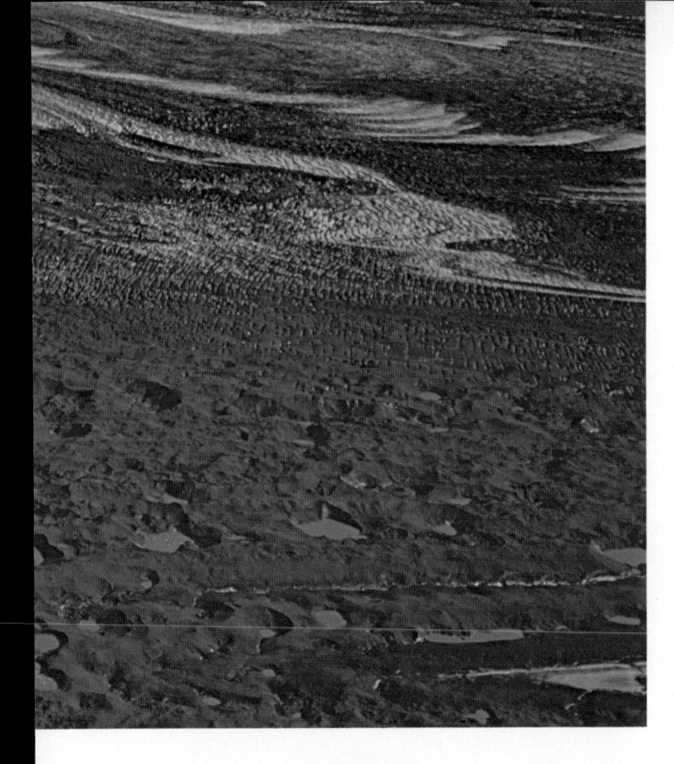

Below — *Melting of stagnant ice once separated from Fairweather Glacier has caused this expanding kettle to form, collapsing a forest that developed on the ablation moraine.*

number of tributary glaciers that have joined to form the large glacier system.

Much of the sedimentary material eroded by glaciers is transported by running water and deposited in layers in front of the glacier on a broad, low-angle surface known as an **outwash plain.** Occasionally a block of ice will be left behind during retreat or a block of ice will be floated by a meltwater stream onto the outwash plain, and then be buried in the sediment. As the ice melts, a depression, called a **kettle,** begins to form, and continues to expand in size until all the ice is melted. An outwash plain with many kettles is termed a pitted outwash plain.

On the outwash plain, the volume of sediment deposited is often so great that it cannot be transported at one time by the quantity of water available. Braided streams, whose meandering channels are filled with ever changing bars, continuously rework the sediment. Examples are the Copper and Alsek rivers in southern Alaska and the Susitna River, south of the Alaska Range. Rock flour is transported in suspension by these braided streams and deposited in large plumes in lakes, bays, Cook Inlet, and the Gulf of Alaska. Valleys leading away from a glacier's front may also receive a substantial amount of outwash fill, known as valley train deposits.

Some sedimentary deposits form where running water comes into direct contact with ice. The most common type of deposit is a poorly sorted sand and gravel mass, called a **kame,** which forms in direct contact with stagnant ice. Kames, which generally have some stratification, form within cracks, holes, or crevasses in the ice or between the ice and the land. face. **Kame terraces,** well stratified deposits, often form between the glacier and its valley wall. When the ice melts, the terrace is left along the valley side and may be confused with a lateral moraine.

Running water on a glacier's surface often washes sediment into crevasses where the sediment remains. When the ice melts, the crevasse fills are left as long, steep, narrow, stratified ridges. Water running under a glacier may erode a meandering channel in the ice that can fill with sediment. As a glacier melts, a winding, stratified sediment ridge called an **esker** may emerge from underneath the retreating terminus. Eskers vary in height from several feet to more than 100 feet, and vary in length from a few hundred feet to several miles.

A delta, a nearly flat, triangular deposit of stratified sediment, forms where glacier meltwater streams empty into lakes or bays. When the glacier's terminus ends in a body of water, blocks of ice break off and float away, a process known as **calving.** If sediment is included in the icebergs that drift away, that sediment is referred to as being ice-rafted away from the glacier.

As the iceberg melts or turns over, the sediment it carries falls to the bottom of the lake, bay, fiord, or ocean in which it is floating. During the Pleistocene such ice-rafted sediment was transported from Alaska and deposited in the Pacific Ocean more than 1,000 miles south of the present Gulf of Alaska shoreline. Today this material is recovered in piston and drill cores of the Deep Sea Drilling Project.

Types of Glaciers

Glaciers come in all shapes and sizes but are usually divided into six general categories: alpine, valley, and piedmont glaciers; ice fields, icecaps, and ice sheets. Today all but ice sheets exist in Alaska. Alpine (mountain and cirque) glaciers generally begin high on the slopes of a mountain or plateau. Valley glaciers originate from ice accumulating in one or more basins that are cut into the mountain or plateau, or overflow from an icecap on top of a plateau. Valley glaciers erode U-shaped channels in which they flow. Many of the larger valley glaciers in Alaska exceed 20 miles in length; at 92 miles, Hubbard, the longest, calves into Disenchantment Bay.

Piedmont glaciers or piedmont bulb glaciers form where two or more coalesced valley glaciers flow from a confined valley and form a broad fan or lobe-shaped ice mass at the base of the mountains. The most spectacular examples in Alaska are Malaspina and Bering glaciers, with piedmont lobes covering more than 800 square miles each.

Icecaps develop where snow and ice fill a basin or cover a low-relief plateau to a substantial depth. When thicknesses become great enough, tongues of ice

Cirque, also called alpine or mountain, glaciers head high on the slopes of a mountain. This cirque, surrounded by a well-developed arête *ridge, is located in the Wrangell Mountains.* (George Herben)

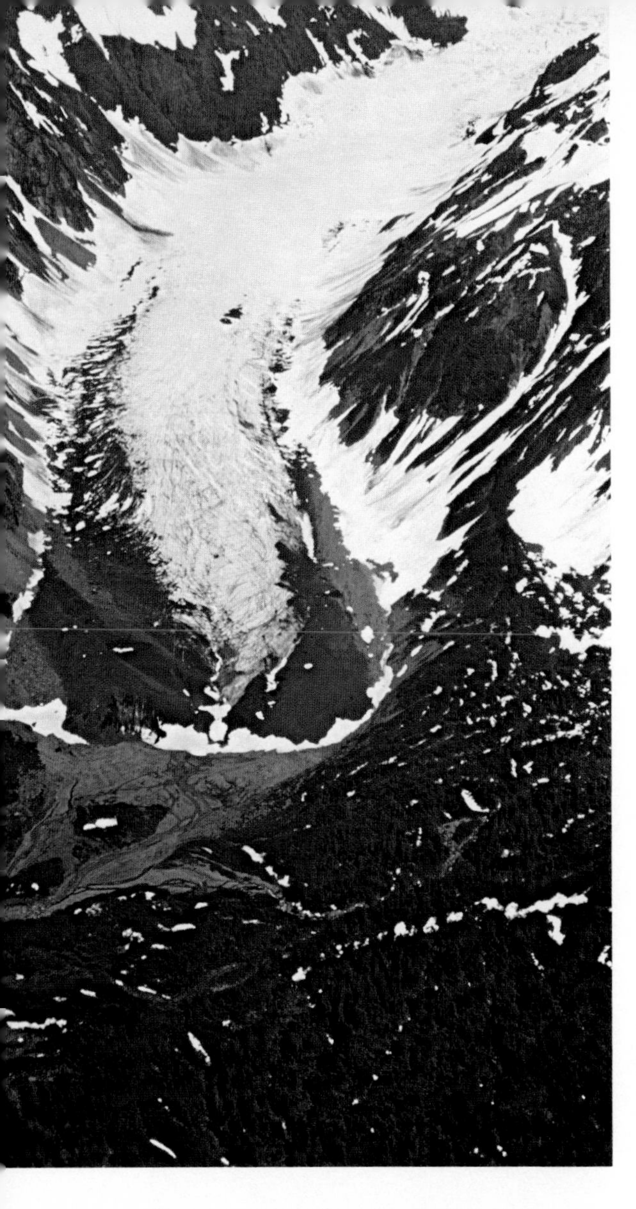

overflow the basins or plateaus as valley glaciers. Ice fields develop where large, interconnecting valley glaciers are separated, in part, by mountain peaks and ridges which project through the ice as **nunataks.** Examples of ice fields in Alaska are Juneau, Stikine, Harding, and Sargent icefields, each of which occupies between 500 and 1,500 square miles.

Ice sheets or continental glaciers are vast ice masses that completely cover a huge land mass. Leading present-day examples are the Antarctic Ice Sheet and the Greenland Ice Sheet. The Antarctic Ice

Above — *This small unnamed valley glacier descends toward Russell Fiord, east of Yakutat. Valley glaciers erode U-shaped channels in which they flow. The longest valley glacier in Alaska is Hubbard Glacier, which flows for 92 miles.*
Right — *Ice fields are areas where large interconnecting valley glaciers are separated by mountain peaks and ridges, which project through the ice. This aerial photo shows the Stikine Icefield, northeast of Petersburg.*

Sheet covers more than five million square miles and in places exceeds 14,000 feet in thickness. Alaska has not had continental glaciers for the last 10,000 to 15,000 years, but during the Pleistocene much of southern and southeastern Alaska was covered by a large continental glacier.

Rock glaciers consist of a valley-filling accumulation of angular rock blocks that resembles a glacier in shape, heads in a cirque, but has little or no visible ice at the surface. Investigations have shown that ice fills the spaces between rock blocks and that rock glaciers move, although very slowly.

A reconstituted glacier, also known as a reconstructed glacier or a **glacier remanié**, forms by the rejoining through pressure melting (regelation) of ice blocks that accumulate at the base of a slope through avalanching or icefalls. Once reconstituted, these glaciers behave like normal alpine or valley glaciers.

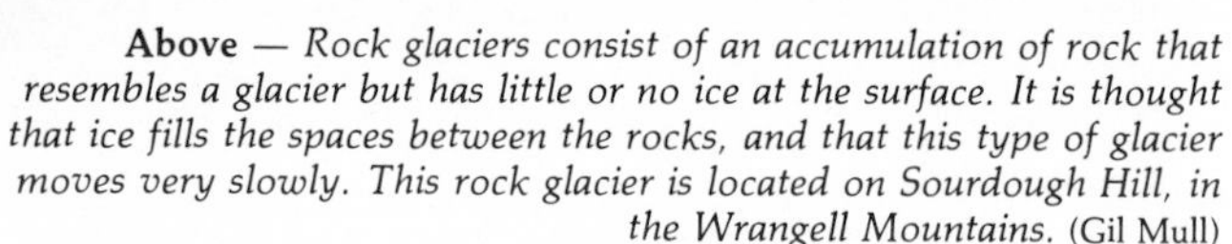

Above — *Rock glaciers consist of an accumulation of rock that resembles a glacier but has little or no ice at the surface. It is thought that ice fills the spaces between the rocks, and that this type of glacier moves very slowly. This rock glacier is located on Sourdough Hill, in the Wrangell Mountains.* (Gil Mull)

Right — *Icecaps develop where snow and ice fill a basin or plateau. When the ice mass becomes thick enough, tongues of ice overflow as valley glaciers. This photo shows the upper portion of Mount Wrangell, which supports a mammoth icecap that flows down in all directions. In places where the ice meets a cliff, it plunges over in dramatic icefalls.* (Ransom Saltmarch)

Opposite — *Cascade Glacier, located in the Chugach Mountains near College Fiord in Prince William Sound, is a typical piedmont glacier. This type of glacier forms where two or more valley glaciers merge and flow from a confined valley to form a broad fan or lobe-shaped ice mass at the base of the mountains.* (Steve McCutcheon)

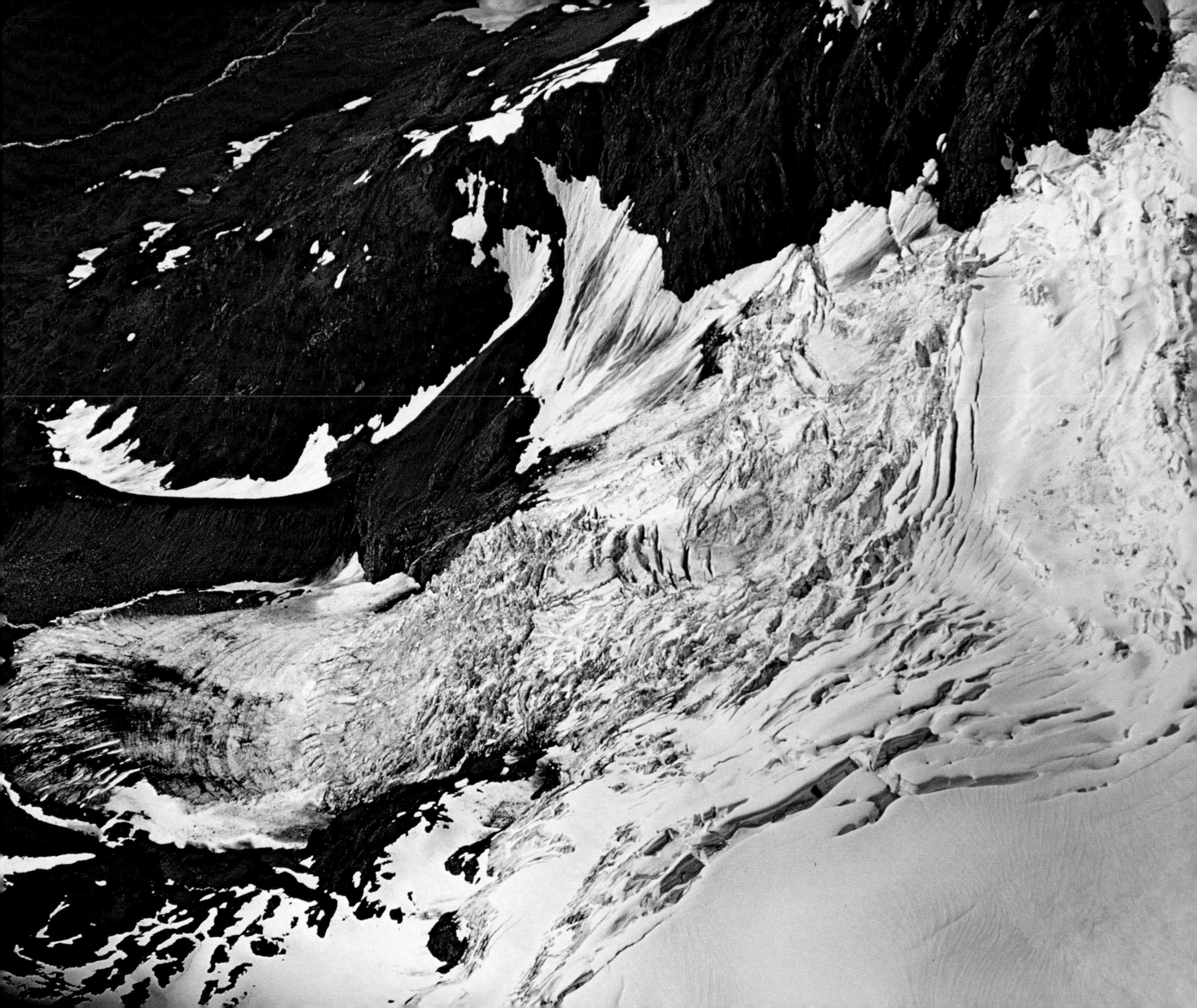

Measurement of Glacier Movement

Studies of glacier movement involve measurement of two completely different aspects of the glacier system. First is the determination of changes in the position of the ice terminus; second is the determination of the rate of ice flow at the glacier's surface.

Rate of ice flow at the glacier's surface is usually determined by a variety of surveying techniques that either involve measuring distances between a point on the surface and a fixed survey point, or by measuring angles between fixed survey points and stakes or flags placed on the ice surface. Normally measurements have to be made over a period of days to accurately determine movement rate.

In the last decade new techniques, using a Doppler laser system, have reduced the time needed to ascertain surface velocity at a specific point to only several seconds.

One method of determining the rate of glacier movement is by comparing photographs taken from common points. Note the changes in the south side of the terminus of Mendenhall Glacier in these two photos, taken in 1974 (left) and 1980 (right). Both were taken from a similar spot on the East Glacier Trail, and show that the glacier had retreated in the intervening six years. Note also that the moraine present in the 1974 photo had disappeared by 1980.

For example, only 50 seconds are needed to accurately determine the velocity of a glacier that moves at a rate of five feet per year. Measurement of rates of movements as small as .000001 inches per second can be made at distances of more than a mile.

Changes in the position of glacier termini have been determined historically by comparisons of photographs taken from common points or in recent years by comparison of aerial photographs. In the last decade two new tools have enabled these measurements to be done by remote sensing. The first is ERTS (Earth Resources Technology Satellite); later the LANDSAT (Land Satellite) program sent polar-orbiting satellites circling over Alaska on a regular basis and transmitting digital photographic information back to earth. Although the resolution (the minimum size object clearly defined from space by a satellite) is only 100 to 200 feet, large changes such as surges and glacier lake drainings can easily be measured.

The second new development, K- and X-band side-looking radar, permit measurement and depiction of surface features on the earth regardless of weather conditions and at distances of up to 50 miles. Other airborne radar techniques permit measurement of glacier thickness and shape.

Above — *These two photos of Variegated Glacier, near Yakutat, are good examples of the value of comparison photography in determining glacial movement. The glacier was beginning to surge in the photo on the left, taken in August, 1964. By the time the photo on the right was taken, precisely one year later, the glacier had advanced approximately four miles.* (Austin Post, U.S.G.S.)

Left — *Douglas K. Bingham measures glacier flow by surveying from the rim of Mount Wrangell's summit caldera in this 1965 photo. The snow-free surface at this point consists mostly of red clay altered from the underlying andesite rock by hydrothermal energy. A rock plug, covered with rime ice, protrudes behind Bingham. Mount Blackburn (16,390 feet) rises on the horizon between the rock plug and exposed clay slope.* (Dr. Carl Benson)

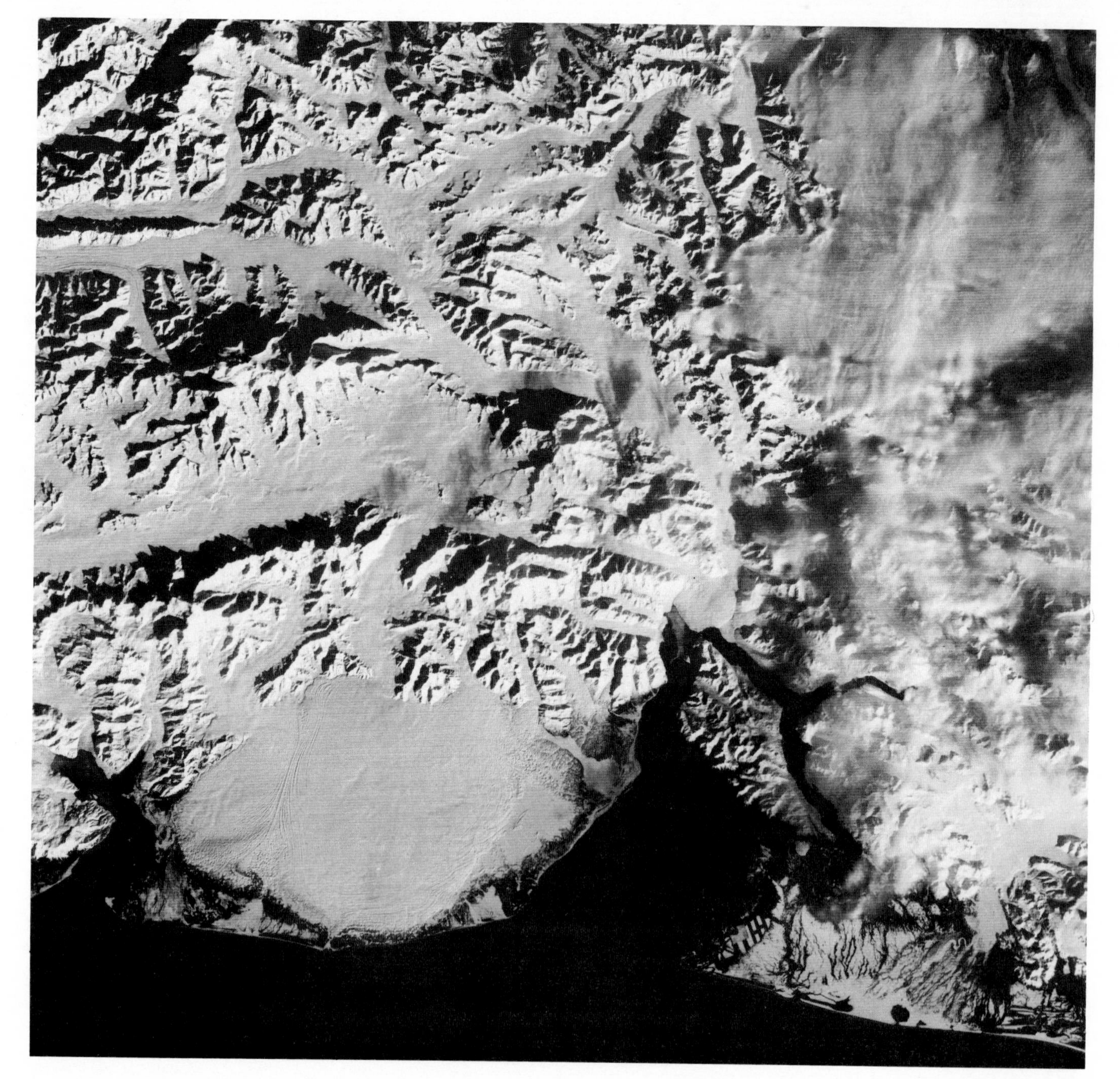

These two high-altitude images of Malaspina Glacier are examples of how modern technology can assist glacial research. At the right is a LANDSAT Image — topographical information is transmitted from an orbiting satellite to a computer, which translates the information into various shades of gray, creating an image such as this. It is not a photographic process. The color photograph is a NASA U-2 Image, taken from an aircraft flying at an altitude of about 63,000 feet. Shot using infrared film, the result is a false-color photograph: vegetation shows up in red and water and ice appear in varying shades of blue.
(LANDSAT Satellite Image 1581-20012 and NASA U-2 Image 4929, courtesy of Bruce Molnia)

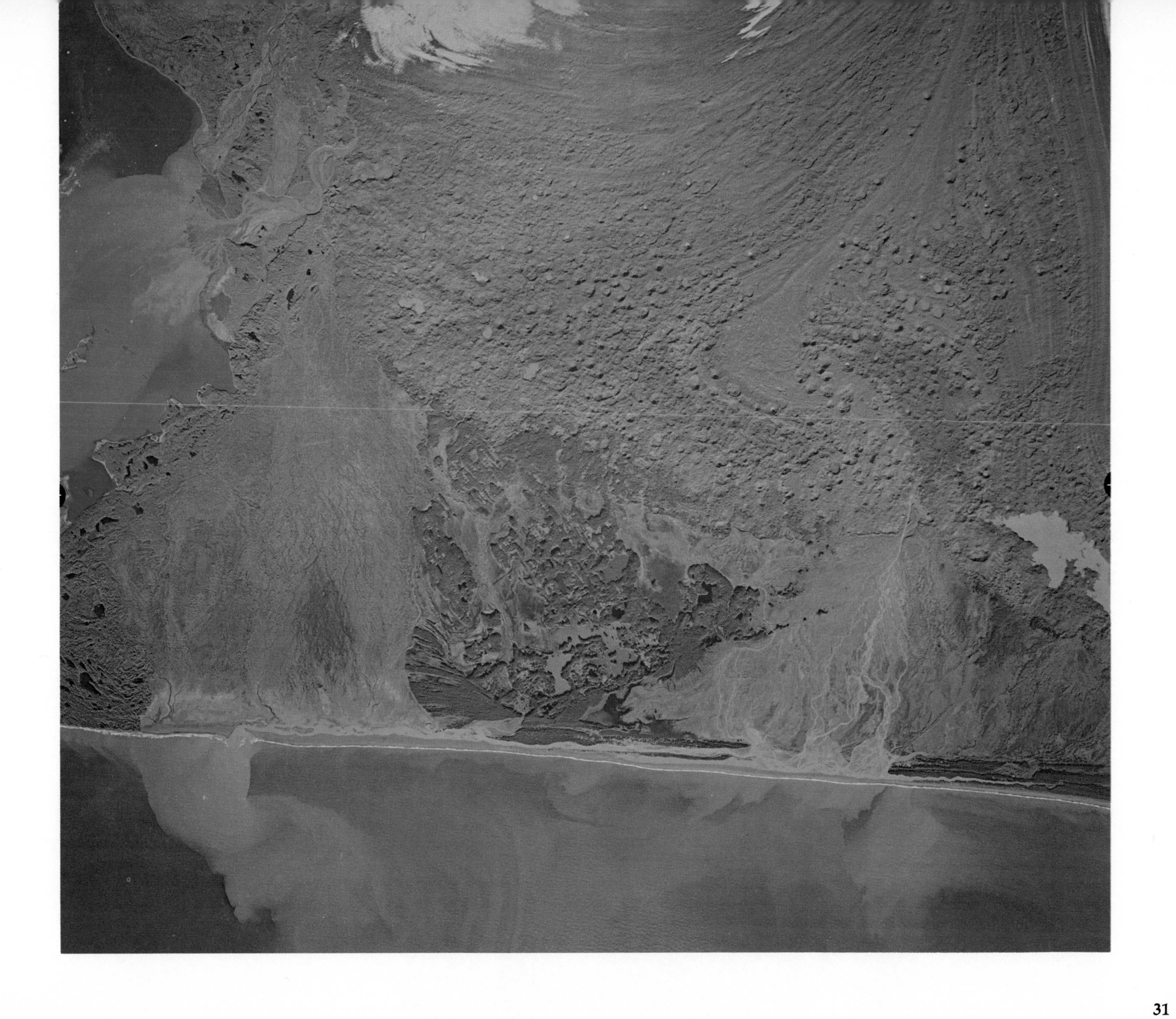

Glaciers in Space and Time

About 300,000 square miles of Alaska are presently covered by active, moving glaciers. Almost 5% of Alaska's total area, this glaciated terrain represents geographic areas from the Canadian border in the southeast, to the Brooks Range in the north, to the Aleutian Islands in the southwest. The greatest concentration of glaciers occurs around the perimeter of the Gulf of Alaska in the Saint Elias and Chugach mountains, and in the Wrangell Mountains and Alaska Range to the north.

During the **Pleistocene**, a period of geological time popularly called the Great Ice Age, more than 50% of Alaska, and part of the continental shelf surrounding the state, were covered by continental ice sheets. A huge ice sheet covered the northern Gulf of Alaska, burying Middleton Island, and this ice sheet may have had a floating, iceberg-calving terminus similar to the present Ross Ice Shelf in Antarctica.

About 10,000 to 15,000 years ago Alaska's massive ice sheets began to thin and melt. Many of the strongholds of glaciation that exist today have endured since the time of these gigantic Pleistocene ice sheets. The ground in unglaciated parts of Alaska was perennially frozen; this condition still persists and is known as permafrost.

Unlike the northeastern United States, where continental glaciers occurred only during the last two million years, Alaska has possessed the right combination of climate and geology to support almost continuous glaciation for at least the last 12.5 million years.

Probably an even older extensive glacier history has existed in Alaska, but the record has been obscured and covered by more recent intensive glaciations.

The last phase of the Pleistocene, called the Wisconsin Glaciation, ended between 10,000 and 15,000 years ago. Climate following deglaciation after the Wisconsin phase has not been uniformly warming; rather, a series of climatic oscillations have occurred.

During the **Holocene,** the name of the period of time following the Pleistocene, there have been four separate episodes of cooling and glacier re-advance. The last, called the **Little Ice Age** (1500 to 1920), saw an increase in the size of many of Alaska's glaciers, many of which are only now retreating from the maximum positions attained during that period. Other, similar periods of glacier advance occurred 7,000 to 8,200 years ago; 4,900 to 5,300 years ago; and 2,400 to 3,300

Presently, about 30,000 square miles of Alaska is covered by glaciers. This satellite image of the Wrangell Mountains shows one of the most heavily glaciated areas in the state. Many glaciers are clearly visible; the largest, at top center, is Nabesna Glacier. (LANDSAT Satellite Image 1422-20212 18 September 73)

years ago. The collective series of glacier advances is referred to as the **Neoglaciation.** During non-glacial time periods, climate was significantly warmer.

About 5,300 to 6,600 years ago, climate was warmer than present, perhaps by as much as 3° to 5° F. This interval during which glaciers retreated throughout Alaska is referred to as the **Altithermal**, also informally called the Thermal Maximum or Climatic Optimum.

The future of glaciers and glaciation stimulates much speculation but is, of course, unknown. Only minor increases in precipitation and changes in temperature are necessary to cause significant buildups in the snow packs of existing Alaska glaciers. This could easily produce the beginnings of a new major advance. Warming and decreased precipitation, on the other hand, could trigger the shrinking of many of today's glaciers.

During the Great Ice Age, which began about two million years ago and ended 10,000 to 15,000 years ago, more than 50% of Alaska was covered by continental ice sheets. Many of the glaciated areas in the state probably look much the same now as they did then, touched only by the forces of nature. (Vincent McClelland)

Glaciers and Ice Age Alaska

Text and photos by Dr. R. Dale Guthrie and Mary Lee Guthrie

Editor's note: *Dr. R. Dale Guthrie and his wife, Mary Lee, live in Fairbanks, Alaska, where Dr. Guthrie, a leading authority on Ice Age Alaska, is a professor in the biology department at the University of Alaska. Mary Lee, a writer and artist, works at home.*

Glaciers fascinate. Standing near a glacier, hearing and watching it move and change, one can easily imagine that the bit of glacier ice chipped off and now melting in a drinking cup was perhaps new snow the same winter hunters painted a mural in a dimly lit cave in Lascaux, France, to celebrate and honor their mammoth kill.

In Alaska, where so many glaciers still exist and where the effects of extensive Pleistocene glaciation are clear to see in a landscape unadapted to modern human requirements, the Ice Age seems only in retreat; to be waiting in the farthest mountains where herds of mammoths and other Pleistocene animals still graze quietly just beyond the horizon. Ghosts, all ghosts, but barely. We have missed the Pleistocene by a mere wink of geologic time. It ended some 10,000 to 15,000 years ago.

Today the Ice Age, like much of yesterday's science, is something of a children's tale. Every parent knows enough of the story to amaze his child with wooly mammoths and saber-toothed tigers, and with great seas of ice that move over continents. It is easy to forget that the outlines of the story have been clear for just a hundred years. The Ice Age was "discovered" in the middle of the 1800s — in great, great granddad's time.

Geology was a young science in the first half of the 19th century, still a branch of natural philosophy. Early geologists were often clergy or engineers, country gentlemen and doctors, members of the voyages of exploration and trade and, sometimes, professors. Most of the earliest inquiries focused on the structure of rock and on the fossils contained in various strata. Surface features — soil covers of loess, sand and clay; sorted and unsorted gravels; and boulders — also attracted attention, particularly large, rounded

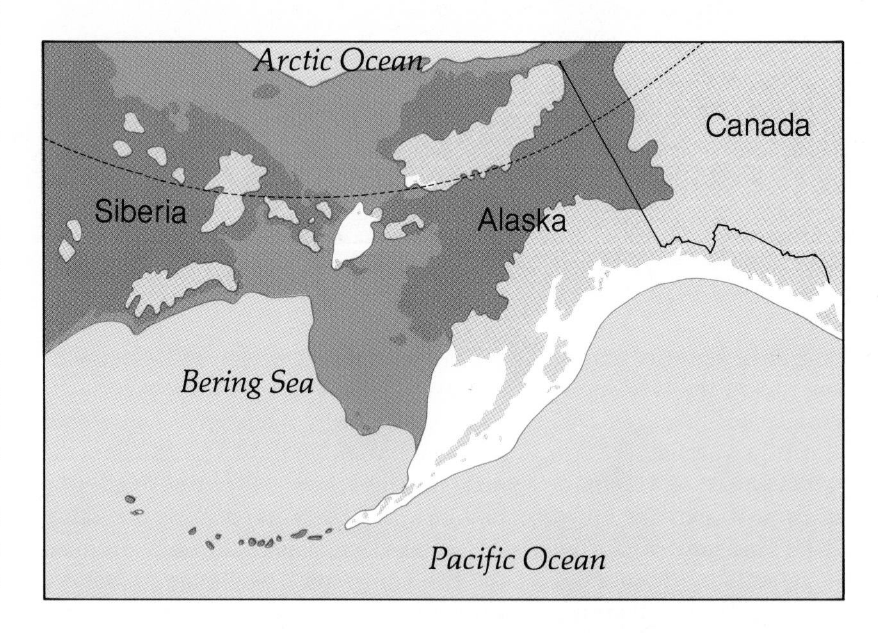

Figure 6 — *This map of Alaska shows the extent of glaciation during the Pleistocene, the Great Ice Age. Source: modified from U.S. Geological Survey.*

Erosion along a sea cliff at Cape Blossom on the Baldwin Peninsula in western Alaska exposes ice wedges, frozen Pleistocene silts and fossils. Much of the Arctic is underlain with ground ice, most of it in the form of ice wedges. Super-cold winter temperatures cause small cracks to open in the ground, allowing spring meltwater to enter and freeze.

boulders of rock that did not match local bedrock in color or composition. These erratics were sometimes enormous — the size of a house — and occurred across Europe and in North America where they were variously explained by local stories as devil's stones, giant's stones, or rubble left by comets. One achievement of early geological enthusiasts was to trace the origins of these erratics to the parent bedrock — sometimes hundreds of miles distant — and to tackle the question posed in this success: how could such rocks be transported so far?

Water, water in the form of a Great Flood, with some assistance from modern geologic theory, seemed the logical agent for working and transporting the erratics and drift deposits the geologists were finding.

Eventually the geologists' work began to produce its own questions. For a time icebergs were incorporated into the flood theory to explain some characteristic grooves on rocks and to account for transport of certain rock material, but the flood theory had a number of difficulties and significant internal criticism when Louis Agassiz and other geologists working in the Alps first clearly associated certain geologic features with glacial action. In the 1830s and 1840s Agassiz and others conducted the first studies on glaciers, measuring ice thickness and movement and observing the ways that glaciers were actually producing striations, polishing rock, transporting huge boulders, leaving moraines, and affecting other geologic features that were impossible to attribute to the action of water.

Agassiz was quick to publish and publicize the new Glacial Theory, first in Switzerland in 1837, and most successfully in Boston in 1846. The response varied from shock and outrage to keen enthusiasm. Agassiz was not the only geologist in the limelight on this issue, but he was a most influential figure, particularly in America where he stayed on to become a professor at Harvard in 1848. For years Agassiz and his students explored New England and west into what was then largely wilderness and Indian territory, tracing the path of glaciation in North America.

By the time Agassiz died in 1873, the Glacial Theory was largely accepted. By 1889 a popular book, *The Ice Age In North America and Its Bearing Upon The Antiquity of Man* by George Frederick Wright, could outline the extent of the ice sheet across the continent, discuss the major features of glacial erosion and transportation, speculate with considerable sophistication on the probable cause of the ice buildup and melting, discuss problems in plant and animal distributions and extinctions, and cite evidence of human occupation in America at the time of the glacial retreat.

Today years of study have made the story more complex. The Ice Age is seen as the culmination of a general cooling during the last 30 million years, of which the most recent three million, the Pleistocene, are characterized by repeated cool and warm periods. These glaciated and unglaciated episodes were tens of thousands of years in magnitude. Once it was thought that four major cycles occurred, but with new data from Greenland ice cores and deep-sea sediments, a long and detailed record of temperatures has documented warm interstadials in what were once thought of as continuously cold periods.

The Pleistocene has been a time of massive and recurrent changes. Ice covered 30% of the world's land mass during glacial maximums in comparison with 10% at present. This resulted in lowered sea levels — as much as 300 feet — and exposed large areas of shallow continental shelf. The weight of enormous ice sheets produced a downwarping of the earth's crust and in places the ground is still rebounding after 11,000 years. Rebound in some areas of the Baltic is lowering the shoreline almost half an inch per year, producing changes clearly visible in a single lifetime. Some Viking and Welsh castles that were built with access to the sea are now landlocked as a result of centuries of such uplift.

Accumulation and melting of the Pleistocene ice triggered changes in animal and plant distributions. Continents which had been isolated for millions of years were connected to other continents by newly exposed land bridges. Mass stirrings of flora and fauna resulted in successful colonizations as well as mass extinctions such as those in South America. Deer, sheep, bison, and mammoths entered the New World from the Old; New World camels and horses invaded the Old World. Each episode of glacial and interglacial change resulted in novel biotic combinations.

The tracks and bones of climate are much more difficult to interpret than those of Ice Age mammals. There is still no definitive understanding of the reasons for glacial and interglacial cycles. The earth does not spin symmetrically on its axis in relation to the sun, and likewise the earth's orbit around the sun is not totally symmetric. These asymmetries produce variations in the amount of radiation received by the earth over long term cycles, and they even compound one another in very long cycles on the order of tens of thousands of years. Although there is still much discussion of the causes underlying the late Tertiary cooling trends, the orbital-rotational cycles seem to be responsible for the cyclic character of Pleistocene glacial-interglacial shifts in the earth's weather pattern. During the glacials, circulation patterns carried moisture to different parts of the earth; stable fronts were fixed in other positions. The sheer presence of the continental ice sheets with the great expanses of highly reflective ice and snow probably generated many climatic changes once they were established: i.e., stable high pressure systems and displacement of weather tracts out around the body of the ice. For example, scientists think that the continental interior of North America received more moisture than at present. Boreal coniferous forests covered the midcontinent to northern Florida. Deserts and grasslands were diffused into less exclusive biomes.

In Alaska thick ice fields covered much of the landscape south of the Alaska Range, but the mountains surrounding the central Alaska interior received most of the precipitation and prevented the accumulation of deep or permanent snow cover in the interior. The Alaskan Refugium, as it is sometimes called, was connected by the exposed Bering Land Bridge to Asia, and Pleistocene Alaska shared many vegetational and faunal characteristics with the cold northern grasslands of Eurasia.

Glacial Silt and Preservation of Fossils

One of the clues of glacial action Agassiz and other early naturalists came to recognize from their study of modern glaciers were certain characteristically rounded and scratched rocks. The slow movement of the glacier as it rolls and abrades rocks beneath the great weight of ice works like a mill, producing fine particles of ground rock. Some of this is washed out away from the glacier by summer meltwater, giving glacial streams their milky appearance. The water is laden with fine rock particles. As the glacial streams braid and shift, these particles are deposited in the outwash fans and river bars where cold air flowing down the glacier catches the sand-sized particles and produces dunes. The smaller silt and clay, however, are picked up and carried by the wind as dust, sometimes for great distances. Much of the unglaciated Alaska interior is covered with deposits of this tan glacial dust called loess. Around Fairbanks these loess deposits are as much as 100 feet thick. But because this dust is easily transported by water, it does not remain in a uniform mantle over the countryside but is washed and flows down slope to form thick deposits in the valley — exposing bedrock along the crests of hills and ridges.

All ground ice is not formed in ice wedges. Some overflow ice from creeks, auf ice, becomes buried by spring silt and remains preserved in the frozen ground. Bedding planes of auf ice, unlike ice wedges, are horizontal and the ice contains very little silt — giving it a blue-white appearance. Here Dr. Troy L. Pewe, one of Alaska's foremost geologists (left), and a friend inspect a layer of auf ice.

In Alaska and Siberia, Pleistocene cold froze these silt deposits in place in the valley bottoms and as new silt was added on top, the underlying silt remained frozen — in many places for tens of thousands of years. As the silt flowed into the valleys, it covered plants and animals, almost like the ash of Pompeii. The remains of organisms buried in the silt continued to rot, but decomposition was slow and was halted prematurely by the frost. Such loess deposits appear darkly organic rather than tan because of their rich load of partly decomposed plants, insects, and mammal remains. Frequently the valley deposits are banded with more stable soils of peat and woody plants, marking warmer interglacial episodes. Layers of buff-colored volcanic ash also occur. These blankets of loess have preserved many clues to plants and animals that occupied the Ice Age landscape.

Preservation of plants and animals is a very haphazard process. Most organisms live and die without a trace. This is particularly true of the smaller herbaceous plants that dominated the glacial vegetation across the north. Animals, especially large mammals with rock-like teeth and fairly durable, large bones and horns or antlers, stand a much better chance of being preserved.

New techniques, many developed since World War II, have added much detail to Pleistocene studies. Today, in addition to large mammal bones, a paleontologist might also study microfossils, tiny teeth and bone fragments of small mammals, insect parts, gastropods, or gastroliths (bird gizzard stones) obtained by washing deposits through fine-meshed screen. New dating techniques are used. Organic and inorganic isotope dating techniques read the proportions of constituent atomic isotopes which change with time to determine the age of the material in question. Linear accelerators are used to date volcanic ash by fission tracks. Climatic data is obtained from cores of deep-sea sediments and Greenland ice sheets. Marine foraminifera provide an interesting twist to the story as the frequency with which they coil in left- and right-handed skeletons reflects changes in sea water temperature.

Pollen analysis has provided many clues to Pleistocene vegetation. Grains of plant pollen are protected by a durable

Layers of volcanic ash interbedded with darker silts can be used in a new dating technique of fission tracking. These beds have been warped and faulted by silt slumping.

waxy coat and because pollen is widely distributed in the wind, fossil pollen incorporated in loess deposits or collected in cores from lake-bed sediments (where it is best preserved) can be examined to determine what plants were present thousands of years ago. The problem with fossil pollen studies in Alaska is that most pollen from glacial periods is herb pollen (from grasses and sedges) and herb pollen is so similar in appearance that it can rarely be identified to genus. Thus we do know that the vegetation in the Alaskan Refugium was herbaceous, but botanists have been engaged in considerable controversy about the details of Pleistocene flora.

Plant parts and plant seeds are also preserved under special conditions: i.e., stomach and gut contents of mummified animals, animal dung in caves, and nests and seed caches of small mammals. In these cases the seeds can sometimes be identified and the plant parts can be carefully processed to obtain cells from the surface of the plant tissue. Cuticle cells are different in each species and they can be used to identify plant tissue that is otherwise unidentifiable.

Although many details are missing about Pleistocene vegetation, some outlines are clear. The cold, dry climate and limited snow cover that characterized the unglaciated interior of Alaska was undoubtedly stressful for plants adapted to keeping much of their mass and nutrients above ground. Under these conditions, plants that could move below ground, herbs and particularly grasses and grass-like species that are able to translocate their nutrients to root systems for winter, would be most successful. Pollen profiles from Alaska lakes show an absence of tree pollen during the glacials; pollen from grass-like plants predominates. There is some evidence that the more stress tolerant balsam poplar *(Populus balsamifera)* did persist in the interior but it is likely that paper birch *(Betula papyrifera)* and the spruces *(Picea)* were totally absent in the far north.

Soils during the Pleistocene were also different. Soils in the Alaska interior were more basic, not nearly as acidic as those today. The annual rain of dust from the glacial outwash fans may have helped replenish soil nutrients. In any case, the Pleistocene steppe vegetation was able to support an impressive variety of large mammals. Some ecologists, such as Valerius Geist at the University of Calgary, have argued that the environment adjacent to glaciers is a very special one for plants and animals. Glacial abra-

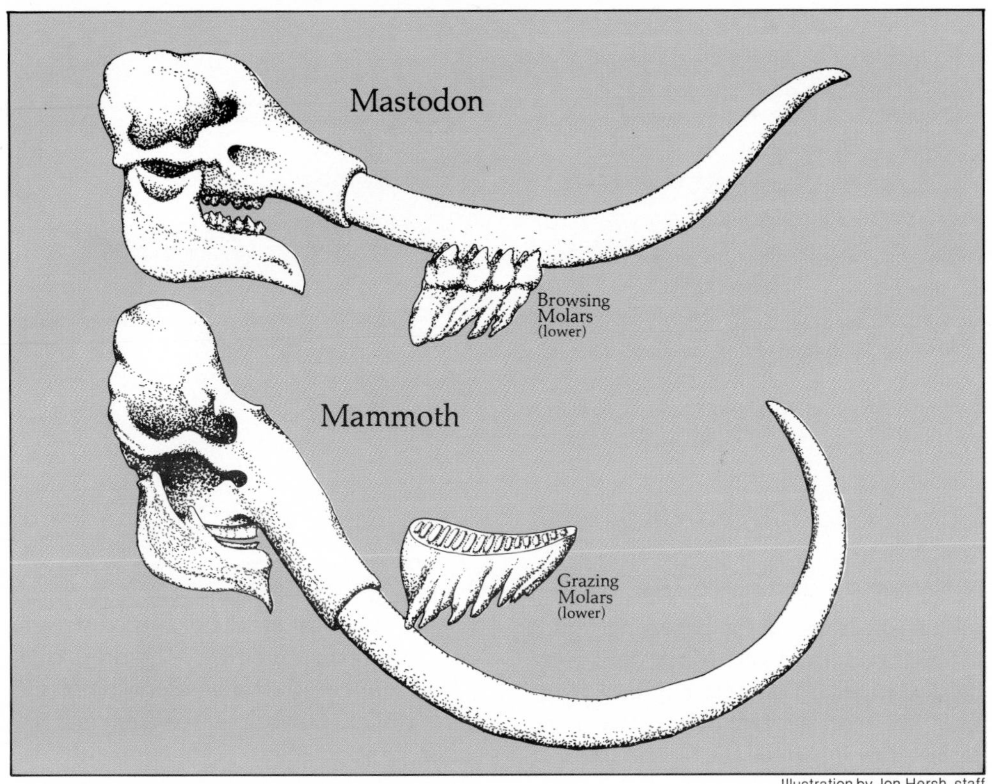

Figure 7 — *Mammoth and mastodon skull and teeth. Throughout North America mastodon fossils are always associated with woodlands, so mastodons were presumably in Alaska during the interglacials since those were the times when trees grew in the far north. Mammoths, on the other hand, were in Alaska at least during the glacials and perhaps during the interglacials as well. Mastodon finds are quite rare; virtually all fossil Alaska proboscidians (mammals of the order* Proboscidea *which have a trunk-like proboscis, such as elephants) are mammoths.*

sion ensures a new supply of mineral nutrients annually and although plant cover may be sparse due to cool summers and long winters, the plants which exist are mineral-rich and well-watered. Such plants provide a high-quality diet for their mammalian herbivores. Animals that inhabit mountainous regions adjacent to modern-day glaciers do tend to be large-bodied and able to expend much of their resources in producing large horns and antlers.

It has been proposed that the same principle could apply to Pleistocene mammals to account for their gigantic size and flamboyant social organs. Pleistocene mammals wore elaborate social garb of long manes, dewlaps, wattles, stripes, humps, tusks, horns, and antlers.

Just as the vegetation patterns of the Pleistocene appear a strange mix, the mammals too appeared in strange places. Musk ox, caribou, and moose roamed south of the ice sheet in what is now Illinois, Iowa, and Tennessee. Likewise mammals from southern biomes lived in the Alaska grasslands: bison, horses, camels, and others. Sheep were common in upland areas. Caribou were present — sometimes fairly numerous — but moose were very uncommon. Some Asiatic colonizers also inhabited the Alaskan Refugium. Saiga antelope, the bulbous-nosed little antelope which now live in the southern Siberian steppe, and Yaks, the hairy relative of cattle and found now only in the mountains of central Asia, both occur as fossils in Alaska deposits. There was a greater diversity of carnivores in Alaska during the Pleistocene. Lions (similar to modern African lions), saber-toothed tigers, giant short-faced bears, ferrets, badgers, and the more familiar northern wolves and grizzlies all lived in Alaska.

Bones of mammals are easily identified to species, so the fossil record provides clear evidence that these strange mixtures of species did once live together. But bones alone do not explain how this was possible. The fact that most of the fossil animals were grazers points to a grassland vegetation. (During the Pleistocene there were few browsers, whereas today the landscape is dominated by browsing moose and caribou.) It is probable that the grazing species each specialized on grasses of differing heights and forbes in different habitats — in a manner similar to that of the animals of the African savanna today. Light snowfall and winter winds probably kept the land free of deep snows, allowing the animals to graze on dead grasses throughout the winter. Mammoth, bison, and horse make up much of the large mammal biomass of the Pleistocene steppe. All three were good grazers, well-adapted to a diet of poor quality grass roughage. In fact, scientists have taken the mammoth, the animal symbolic of the Ice Age and the characteristic inhabitant of the cold grasslands that reached from Alaska to England, to name this Pleistocene biome the Mammoth Steppe.

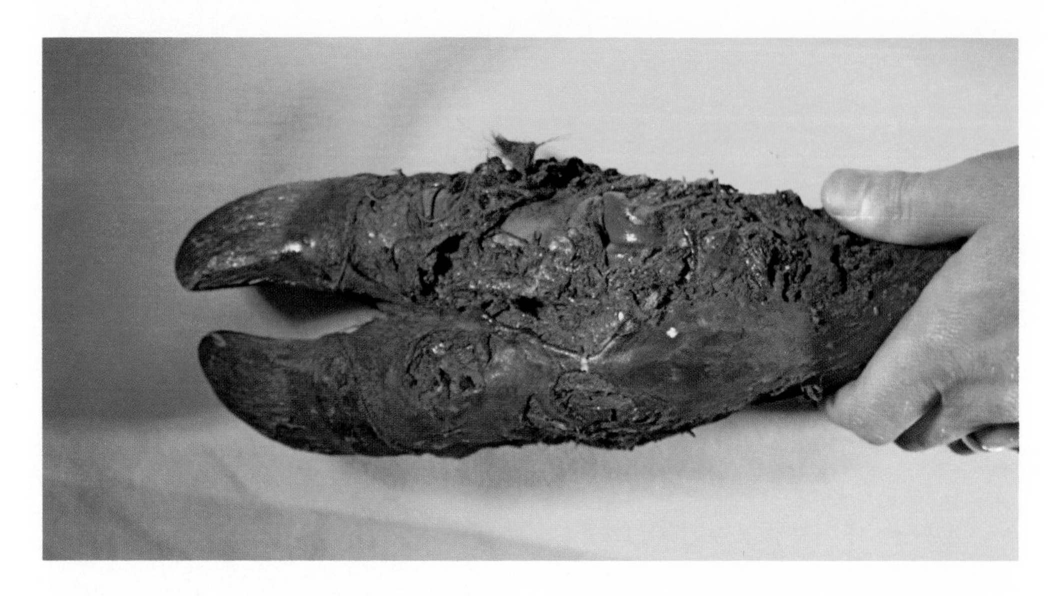

Mammoths, Mastodons, and Mummies

Of all extinct Pleistocene mammals, the wooly mammoth is by far the best known. Mammoths were widespread during the Pleistocene and skeletal remains of mammoths and their distantly related cousins — mastodons — have attracted attention for several centuries. Mastodons were also proboscidians, but their appearance, diet, and habits were quite different from the mammoth. Mammoth teeth are multi-ridged and complexly cemented to withstand a grazer's diet of abrasive grasses and herbs. Mastodon teeth are simple, with high conical cusps, designed for browsing and a diet of leaves and bark. Mastodons were slightly smaller than mammoths, averaging less than ten feet high. During the glacials they inhabited the forests south of the ice sheet and on into South America. During some warmer interglacial episodes, mastodons ranged as far north as Alaska.

Mammoths are not the only Pleistocene animals to appear as mummified carcasses in Siberia and Alaska. Pleistocene bison, horses, musk ox, and other species have also been found. Elsewhere, carcasses are occasionally preserved by means other than cold: dry caves in the American southwest, oil and salt seeps, and limestone caves sometimes yield mummified animals.

But mummies are not found in glaciers or in ice wedges. They are frozen in silt, in muck beds, and if there is ice associated with the mummy it is ice from water seeping into frost cracks or buried stream ice or stream overflow. During cold glacial winters, the valley deposits of silt were chilled far below freezing and as water contracted at such temperatures, the wet soil cracked like a drying mud pan. These cracks were thin, perhaps a few millimeters wide, but sometimes as deep as several feet. From an aerial view the cracks formed large polygonal patterns, as much as 20 feet across. In the spring, before the ground had time to warm and expand back to its summer dimensions, water flowed into the cracks and froze. Year after year this process was repeated and the tiny wedge of ice expanded, pushing the soil toward the center of the polygon. Over time these ice wedges became enormous, building to the point that the mass of ice in the polygonal web was almost as great as the soil in the centers. These were the ice structures the first explorers to the north saw and inter-

Left — *A mummified front foot of the extinct Alaska steppe bison* (Bison priscus) *was found near Fairbanks.*
Below — *A view of the northern hemisphere from the North Pole during the peak of the last glaciation, approximately 18,000 years ago. The black areas indicate the extent of the Mammoth Steppe, defined as the distribution of wooly mammoth* (Mammuthus primigenius), *steppe bison* (Bison priscus), *and northern horses* (Equus *sp.*), *the dominant large mammal grazers of the northern Pleistocene grasslands.*

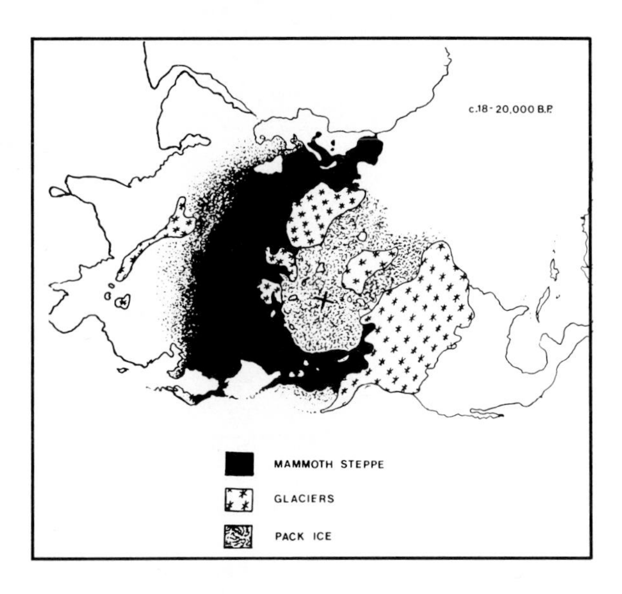

preted as icebergs embedded in the ground. And it is these ice wedges that also gave rise to the image of frozen mammoths melting out of great cakes of ice.

On Alaska's North Slope such ice wedges are forming today, producing the polygonal ground pattern which is so common there. In fact, much of unglaciated Alaska is underlain with Pleistocene ice and frozen ground — and the millions of fossil mammals preserved therein. The vegetation cover can be misleading and permafrost can occur in surprising places — unsuspected until a home or road built above begins to shift and settle slowly out of sight as the ground melts.

Travelers can view this underground landscape along creek banks where erosion exposes the ice wedges from side view and thaws the dark muck. Bones of mammals high grade out of the sloughing mud and are deposited on gravel bars, often to be discovered by canoeists and kayakers. The earliest accounts of Alaska mammoths come from the coast near Kotzebue Sound where frozen muck and ice deposits were exposed by sea erosion. But some of the most productive exposures are not natural but were directly caused by men in search of that precious metal — gold.

Gold Miners and Paleontologists

Mining activity in Alaska has uncovered most of the Pleistocene fossils and mummies that now comprise our museum collections. During the early Pleistocene, gold eroded from exposed schist and was carried down the valleys to be buried with old Pleistocene gravels just above bedrock. Reworked glacial silt washing down from the hills built up the thick muck deposits that are so rich in fossils and such an impediment to gold miners.

Early miners in Alaska sank shafts through frozen muck and gravel to reach gold deposits. Later when large mining companies moved in and bought up many smaller claims, dredges were constructed to process hundreds of tons of gravel a day, but first the frozen silt overburden had to be removed. The insulating cover of vegetation was stripped and the silt allowed to thaw, accelerated by enormous jets of water which also washed the silt downstream. Ice wedges, peat layers, and fossils were thus exposed over hundreds of acres.

In the 1920s an expedition under the leadership of Dr. O. Kaisen from the American Museum of Natural History in New York City collected some bones. When Dr. Kaisen returned to New York, he told other paleontologists about the wealth of fossils being exposed in the mining areas in Alaska. One of these paleontologists, Childs Frick, a member of the wealthy Frick family, was already engaged in collecting fossils from all over the world. Frick contacted the University of Alaska and arranged to have these newly exposed Pleistocene fossils collected from mining areas and included in his personal collection. Otto Geist, then a young German immigrant, was hired to do the actual work of collecting the fossils and as a result much of the material exposed by gold mining activity was saved. The Frick collections were later donated to the Museum of Natural History in New York where they now reside. Otto Geist continued to work in Alaska

A placer miner's "Giant" — a powerful jet of water — washes thawed Pleistocene silt to expose the gold-bearing gravels below. Thousands of Ice Age mammal bones have been unearthed by this process.

for more than 30 years. Essentially a self-trained scientist, he was one of the most outstanding and colorful figures in early Alaska paleontology and archaeology.

The First Alaskans

This review of Ice Age Alaska has thus far neglected the story of one of the most interesting Pleistocene species — human beings. Indirectly, the climatic events of the Ice Age were critical to human evolution. For millions of years in the Late Tertiary our line was an insignificant primate, a background character in the major dramas in plant and animal evolution. But something happened in the unstable stresses of the Pleistocene to spur human development.

Early people, heavy browed and muscled, colonized Europe and Asia from Africa more than half a million years ago. As the Pleistocene unfolded, these early peoples left a fairly good record of their existence in Africa and Eurasia. Generally these human ancestors moved farther north during the interglacials and retreated southward with successive ice advances. By 35,000 years ago people, very similar to humans today, had developed from the ancestral neanderthals. Late Pleistocene animals — gigantic mammoths, wooly rhinos, saber-toothed tigers, herds of horses and bison — all were hunted by men. The images of the animals they painted and carved are a powerful link to the world these first men saw and to their fully human status.

The record of human habitation and hunting is much older in Europe and Asia than in the New World. There seems to have been a barrier to colonization of the far north and thus to movement into Alaska and the rest of the New World until around 13,000 years ago. There are no sites of the late glacial people north of 60° in Asia that date earlier than 13,000 years B.P. It appears that people first discovered new ways to cope with and exploit the far north about that date. Major climatic changes in seasonality were taking place at this time — deeper winter snows and tree lines rapidly moving northward — which also may have contributed to the new human colonization. The Berelekh Site, a mammoth hunters' camp in northern Siberia, dates at 11,000 to 12,000 B.P. So does the Dry Creek Camp Site excavated just north of the Denali National Park and Preserve entrance. A number of other archaeological sites have been found in Alaska which date in this same range. Evidence for earlier human habitation in the New World is very sketchy and is the focus of current dispute.

Once northeastern Asia and Alaska were reached, access to North America was guaranteed — and life in the New World would never be the same. In any event, there are many sites scattered below the southern margin of the North American ice sheet that date in the 10,500 to 11,500 B.P. range. All of these early sites are identified by a common style of stone spear point and lithic technology called "Clovis" after the New Mexico town where such points were first described.

With the onset of climatic changes at the end of the last glacial, the dry Mammoth Steppe deteriorated; Alaska, like northern Asia, became wetter. We know from pollen studies that shrub tundra began to dominate many areas, starting about 14,000 years B.P. By 10,000 years B.P. the northern tree species common today began to be a significant part of the landscape. These combined climatic/vegetation changes thus seem to account for the extinctions and range restrictions of Pleistocene mammals in Alaska. Alaska vegetation is not now suitable for the large Ice Age grazers; deep snows in many areas now prevent access to the grasses which do exist. Mammoths and horses became extinct about 12,000 B.P. Bison, being more adaptable, managed to survive in several areas of Alaska but they did become smaller in body size. (Bison finds in Alaska are dated as late as 500 years B.P.)

In the post-glacial landscape with its wetter summers and deeper winter snows, browsing caribou and moose — which are well-adapted to these conditions — became the most common large herbivorous mammal species and continue so today.

Humans seem to have fared well during the early part of the post-glacial in interior Alaska, as judged by numerous sites with microblades and wedge-shaped cores of quality stone. But there are long gaps suggesting inhospitable times in the archaeological record of the interior peoples. Yet in coastal Alaska the record of occupation appears to be continuous, even though there have been major cultural changes.

Excavation of the Dry Creek Camp Site, an early man site near Healy, Alaska, dated at about 11,000 years ago. The early peoples who camped here were hunting bison, elk, and sheep.

Thus Alaskans live in a land with many reminders of the Ice Age — there are ties to the Pleistocene at every turn: our silty glacial streams, the rounded boulders with which we circle our campfires, and the dust that rises on country roads, but most especially in the grand masses of ice called glaciers.

Glacial Effects on the Earth's Crust

Eustacy and Isostacy

Eustacy and **isostacy** are two independent processes that are responsible for base level changes in glacial environments. Eustacy refers to the worldwide sea level regime and its fluctuations caused by changes in the quantity of sea water available. Isostacy is a condition within the earth's surface where material within the crust and mantle is displaced in response to the increase or decrease in mass at any point on the earth's surface.

During the Pleistocene the growth of continental ice sheets tied up so much fresh water that the worldwide sea level was lowered about 300 feet. As a result of this eustatic sea level lowering, the continental shelves of the world, not only in glacier-covered areas but everywhere, were exposed. Between 12,000 and 15,000 years ago the major continental ice sheets began to melt and sea level started to rise. Equilibrium, near the present level, was reached only about 4,000 years ago. Melting of the Antarctic and Greenland ice sheets, which contain the majority of all glacier ice, and the polar ice mass, would result in additional eustatic sea level rise of about 300 feet. Further growth of glaciers would cause eustatic sea level lowering.

In Alaska eustatic sea level lowering left many present fiords isolated from the Pacific Ocean. Many glacial features formed when glaciers advanced onto the exposed continental shelf and were covered by rising sea level.

As a result of an increase in mass of material caused by the presence of hundreds of thousands of feet of glacier ice overlying the earth's surface, many areas of Alaska were depressed. Removal of the ice mass has resulted in the slow rising of the ground surface through isostatic readjustment. The Bartlett Cove area of Glacier Bay has risen an average of about one inch per year since it was deglaciated about 200 years ago. This rate of isostatic readjustment is also valid for many other Alaska areas.

A eustatic sea level rise of about 300 feet — which would be the result of melting of the Antarctic and Greenland ice sheets and the polar ice mass — could cause dramatic changes in many of the state's tidewater glaciers. Such a rise could possibly put part of the massive piedmont lobe of Bering Glacier under water. (Gil Mull)

Present Distribution of Alaska's Glaciers

The existence of glaciers is a direct result of a unique combination of climatic and geographic features. For a glacier to exist, the right combination of temperature, elevation, moisture source, and area for accumulation to nurture the growth of a glacier are necessary. Alaska has dozens of geographic regions that possess all the elements required for glacier formation and growth.

Today, glaciers exist in the Coast Mountains of southeastern Alaska; in the Fairweather, Saint Elias, and Alaska ranges; and in the Kenai, Chugach, and Wrangell mountains. North of the Arctic Circle, many small mountain and valley glaciers exist in higher summits of the Brooks Range, especially in the Romanzof Mountains. Many volcanoes in the Aleutian Islands and Alaska Peninsula are glacier-covered. As many as 50 small alpine glaciers remain on the backbone of Kodiak Island and in the Kilbuck and Wood River mountains, south of the Kuskokwim River, and even in the Kigluaik Mountains of the Seward Peninsula.

In all, counting every tiny cirque glacier, valley glacier, and permanent ice mass up to the size of Bering and Malaspina, Alaska may possess more than 100,000 separate glaciers. Of these, less than 1% have been named, and an even smaller percentage have been studied.

Distribution of Alaska's Glaciers	Approximate Area (Square Miles)
Chugach Mountains	8,200
Saint Elias Mountains	5,500
Alaska Range	5,300
Wrangell Mountains	3,200
Coast Mountains	2,800
Kenai Mountains	1,850
Alaska Peninsula	less than 1,000
Aleutian Islands	less than 1,000
Brooks Range	less than 1,000
Kodiak Island	less than 20
Kilbuck and Wood River Mountains	less than 10
Seward Peninsula (Kigluaik Mountains)	less than 10
Total	approximately 29,000 square miles

(Source: Modified from The Alaska Almanac®: Facts About Alaska, *1981 edition)*

The less-than-precise measurement of glacial areas is due to at least three factors: 1) in some remote areas no field observations have ever been conducted; 2) in snow- and ash-covered areas, it is very difficult to determine where glaciers exist or where bedrock is buried by only a thin snow or ash cover; 3) many small glaciers, and even some larger ones, have completely disappeared since the time when first observed.

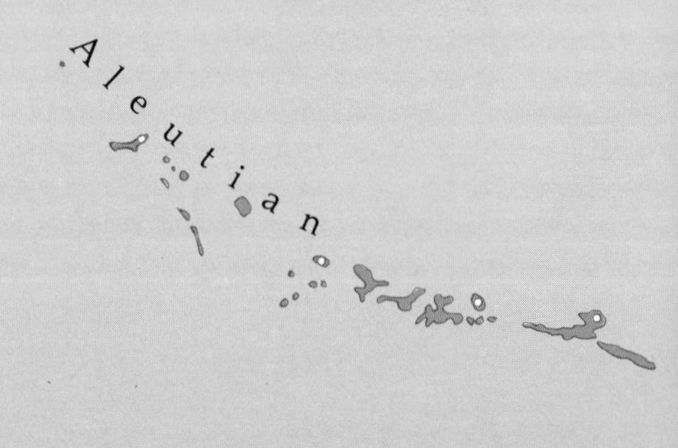

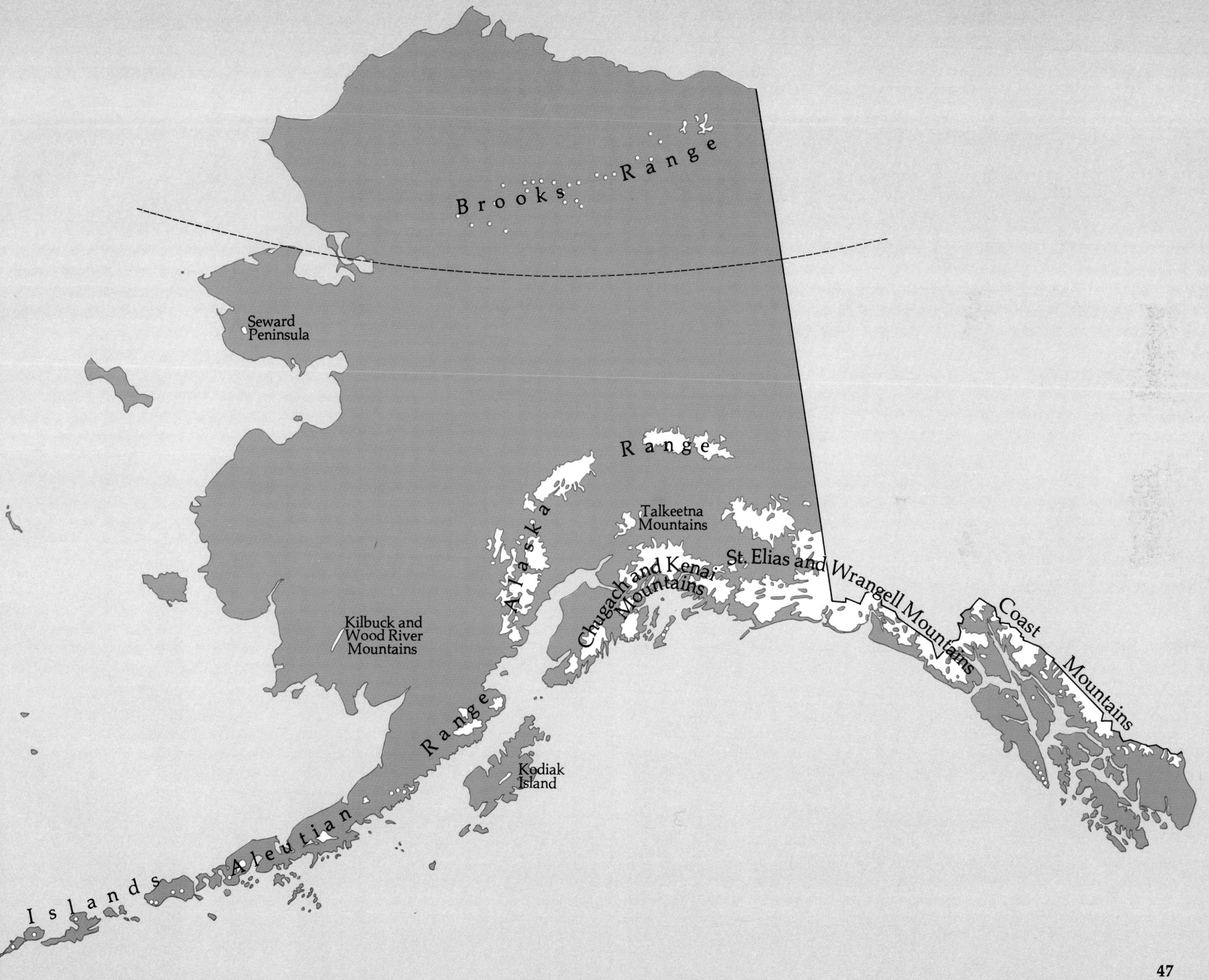
Brooks Range
Seward Peninsula
Alaska Range
Talkeetna Mountains
Chugach and Kenai Mountains
St. Elias and Wrangell Mountains
Coast Mountains
Kilbuck and Wood River Mountains
Kodiak Island
Aleutian Range
Aleutian Islands

Left — *The high peaks of the Chugach Mountains are covered in perpetual snow* (névé) *and give rise to countless glaciers. These mountains, with a glaciated area of approximately 8,200 square miles, contain the highest concentration of glaciers in Alaska.* (Jon R. Nickles)
Below — *Several glaciers can be identified in this photo of the northeast side of the Juneau Icefield, in the Coast Mountains. The icefield's glacier systems are the subject of detailed, long-term study, through the Juneau Icefield Research Program.*
Right — *The Brooks Range supports a large number of small cirque and valley glaciers, such as this unnamed glacier which flows from the south side of Mount Michelson (8,855 feet). The highest concentration of Brooks Range glaciers is found in the Franklin and Romanzof mountains.* (Gil Mull, reprinted from *ALASKA GEOGRAPHIC®*)

Exploration of Alaska's Glaciers

An entire volume could be written about the explorations of Alaska and its glaciers. It is surprising to note the lack of recognition of the presence of glaciers in southern Alaska while reading accounts of 18th century voyages by such astute observers as Malaspina, Vancouver, Puget, and Cook. All of them took for granted that the walls of ice and snow they found were part of the northern polar barrier and not glaciers. In 1778 Cook and Vancouver sailed into Cross Sound past the gigantic glacier that filled the mouth of the Glacier Bay and failed to recognize it. This is remarkable considering Cook had previously been to the Antarctic.

The first recorded observation of a glacier calving in Alaska was made during Vancouver's 1794 exploration of College Fiord in Prince William Sound. A correlation was made between the frightening thunderous roars that were frequently heard and the blocks of frozen snow falling off the faces of "snow cliffs." But the snow cliffs were not recognized as glaciers.

Of the 18th century explorers, only Jean Francois Galaup de La Perouse recognized the significance of glaciers he observed in 1786. La Perouse's map of Lituya Bay, published in 1797, carefully and accurately depicts the presence and location of five named glaciers located in the upper ends of the bay. The accuracy of La Perouse's map has provided glacial geologists with a base from which to compute changes in the positions of glaciers through the years.

Little was written about Alaska's glaciers during the first half of the 19th century. Between 1848 and 1850 a series of charts and maps appeared that were prepared by Terentiev and Kadin for Mikhail Dmitrievich Tebenkov, director of the Russian American Company and governor of Russian America. These maps, published as the *Atlas of the Northwest Coast of America* clearly showed the location of many coastal features, including glaciers.

In 1863 the Russian Naval Squadron invited Professor William P. Blake, an American, to accompany them on an exploration of the Stikine River Valley. Blake's investigation resulted in the 1867 publication of *The Glaciers of Alaska, Russian America,* the first scientific summary about Alaska glaciers. Between the time of Alaska's purchase by the United States in 1867 and World War I much knowledge about the size, location, and activity of Alaska's glaciers resulted from U.S. government surveys, reports of

In 1786, French explorer Jean Francois Galaup de La Perouse sailed into Lituya Bay while in search of a northwest passage. La Perouse was the only one of the 18th century explorers to recognize the significance of the glaciers he observed.
(Alaska Historical Library)

Left — *Ice ax in hand, John Muir leads a friend's dog across the jagged ice of Brady Glacier in southeastern Alaska. For his trip across the glacier, named for Rev. John G. Brady (1848-1921), Alaska missionary and later Territorial Governor from 1897 to 1909, Muir made booties from his handkerchief to protect the dog's feet.* (Painting by H. Tom Hall, ©National Geographic Society, from *John Muir's Wild America.*)

Below — *La Perouse's map of Lituya Bay accurately shows the presence of five glaciers in the upper ends of the bay. The map, which was published in 1797, has provided geologists with a base from which they could determine changes in the positions of Lituya Bay's glaciers.*

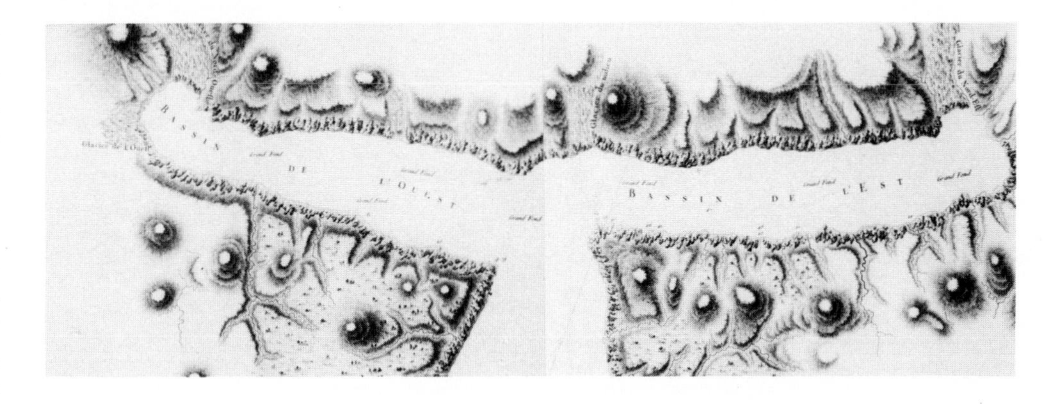

climbing expeditions, and the privately funded observations of a few naturalists and geologists. A few individuals and expeditions involved in these explorations deserve special mention.

John Muir, naturalist and travel writer, visited Alaska for the first time in 1879. Accompanied by Samuel Hall Young, a Presbyterian missionary, Muir explored Glacier Bay and Muir Glacier. In later visits in 1881 and 1899, Muir noted changes in the bay and explored numerous glaciers in other parts of Alaska. Muir's writings were popular and easily available, and may have stimulated others such as Henry Fielding Reid, who mapped Glacier Bay in detail in 1890 and 1892.

Israel C. Russell, a geologist with the U.S. Geological Survey, explored the geology of the Yakutat Bay region in 1890 and 1891. The expedition, jointly funded by the U.S. Geological Survey and the National Geographic Society, had as its primary objective the ascent of 18,008-foot Mount Saint Elias. Although the ascent failed due to severe weather, Russell explored much of Malaspina Glacier, Yakutat Bay, and Russell Fiord, which was later named for him. Russell's expedition produced the earliest photographs of Yakutat Bay, Malaspina Glacier, and the Mount Saint Elias region.

The Harriman Alaska Expedition of 1899, privately funded by Edward Henry Harriman as a vacation for his family and a multi-disciplinary scientific study, was the most prolific scientific exploration of Alaska of its time. Twelve separate volumes of scientific observations were published as a result of the expedition's investigations.

Harriman chartered the steamer *George W. Elder* and invited several distinguished scientists — ethnologists, zoologists, botanists, geographers, and geologists —

Right — *An amateur photographer sets up his camera at Muir Glacier in this 1886 photo taken by W.H. Partridge. Partridge used the new gelatin dry plates, which enabled the photographer to store them after exposure and develop them later in his studio. Prior to the discovery of the dry plate process, photographers had to carry all of their supplies with them, including their darkroom tents.* (Tom Sexton, reprinted from *The ALASKA JOURNAL®*)

Below — *Well known photographer W.H. Partridge caught a steamship playing hide-and-seek with an iceberg in this photo, taken in the mid-1880s. Takon Inlet, mentioned in the title, was probably a variation in spelling of Taku.* (Tom Sexton)

to sail to Alaska. According to the *Harriman Alaska Series, Volume 4* (1910), the expedition sailed from Seattle on May 31, 1899, covered more than 9,000 miles, and made about 50 scientific stops including Annette Island, Wrangell, Juneau, Glacier Bay, Sitka, La Perouse Glacier, Yakutat Bay, Prince William Sound, Homer, Kodiak, Dutch Harbor, Bogoslof Island, Saint Paul Island, Saint Matthew Island, and Cape Fox. The expedition returned to Seattle on July 30, 1899.

Five geologists were aboard the *Elder:* William H. Dall of the Smithsonian Institution and the U.S. Geological Survey, John Muir, Benjamin K. Emerson of Amherst College, Charles Palache of Harvard University, and Grove Karl Gilbert of the U.S. Geological Survey. Edward S. Curtis, later to win fame for his photographs of American Indians, was the expedition's official photographer. Detailed observations in Glacier Bay and Yakutat Bay confirmed and expanded the limits of previous knowledge of glaciers. In Prince William Sound, the expedition explored and named the glaciers of Col-

These photos of Twin Glaciers and Twin Glacier Lake were taken as part of a U.S. Geological Survey aerial survey in 1926. Recent maps of the area, located in the Coast Mountains on the Taku River, show that the glaciers have retreated, considerably enlarging the lake.
(Alaska, 185, U.S.G.S.)

lege Fiord and discovered a completely unknown inlet which they named Harriman Fiord. Two volumes were published summarizing the geological and glacial findings of the expedition.

Ralph Stockman Tarr, of Cornell University, and Lawrence Martin, of the University of Wisconsin, were involved in a series of explorations between 1905 and 1913 that led to more than 50 scientific publications, culminating in 1914 in *Alaskan Glacier Studies,* published by the National Geographic Society. To date this is still the single most comprehensive volume ever published on Alaska's glaciers.

At about the same time that Tarr and Martin were working in Yakutat Bay, Prince William Sound and Kenai Peninsula glaciers were being observed by U.S. Grant and D.F. Higgins of the U.S. Geological Survey. Their results, published in 1913, present the first detailed description of Kenai Peninsula glaciers.

Following World War I, the number of investigations of Alaska's glaciers were significantly decreased while the use of photography increased. Many of Alaska's southeastern glaciers were photographed from the air between 1926 and 1929 in a joint program funded by the U.S. Navy and the U.S. Geological Survey. Much of Alaska was again surveyed by aerial photography in 1941.

William S. Cooper of the University of Minnesota investigated the flora and ecology of Glacier Bay in 1916, 1921, 1929, and 1935. In 1935 he also visited Prince William Sound. These studies documented the cycle by which vegetation first becomes established on bare, recently deglaciated bedrock, and develops through a succession of plants to culminate in mature spruce and hemlock forests.

In 1926 William O. Field of the American Geographical Society began studies of the nature and positions of Alaska's glaciers. His observations, which continued through 1979, comprise the largest existing data base on the changes in Alaska's glaciers. Field established numerous photographic stations that were reoccupied in successive years, thus compiling photo documentation of changes in glacier termini positions. In 1975 the Cold Regions Research and Engineering Laboratory of the U.S. Army Corps of Engineers published William Field's *Mountain Glaciers of the Northern Hemisphere,* a two-volume set complete with atlas. This work is the most complete published summary of northern hemisphere glaciers and the best summary of the history of Alaska's glaciers since the 1914 publication by Tarr and Martin.

In the early 1930s Bradford Washburn began photographic reconnaissance of Alaska mountains as a prelude to many of

his climbing expeditions. His photographs include the first aerial observations of glaciers in the Fairweather, Saint Elias, Chugach, and Alaska ranges.

Following World War II, long-term field programs were established on the Juneau Icefield by William O. Field and Maynard Miller; on the Malaspina-Seward glacier system by Robert Sharp; and at various places in Alaska by Ohio State University's Institute of Polar Studies, the Arctic Institute of North America, and the University of Alaska. In 1955 Richard Hubley of the University of Washington began systematic aerial reconnaissance of Cascade Mountains glaciers. In 1960 Austin Post, sponsored by the University of Washington and funded by the National Science Foundation, expanded this technique to cover most glaciers of western North America. In 1964 Post joined the U.S. Geological Survey and has continued annual photographic observations up to and including 1980. Many of the most spectacular aerial photographs of Alaska's glaciers have resulted from Post's work. In 1970 Larry Mayo of the U.S. Geological Survey began seasonal aerial studies to examine short-term glacier fluctuations. His work is now in its second decade.

The 1964 Alaska earthquake generated a number of scientific studies, most concerned with the effect of avalanches and landslides on the rates of melting and flow of glaciers.

Vegetative studies, similar to the work of William Cooper, but also including

Left — *The terminus of Sheridan Glacier, which drains into the Gulf of Alaska, near the Copper River delta, is still covered with debris from the 1964 Alaska earthquake. In some places, the debris reaches a thickness of six feet.*
Right — *This photo, taken in 1974, shows landslide and avalanche debris from the 1964 Alaska earthquake on a small, unnamed glacier in the Chugach Mountains. The earthquake generated many scientific studies, most of which were concerned with the effect of such debris on the rate of glacier flow and melting.*

long-term climate investigations based on the analyses of pollen profiles augmented by radiocarbon dating, have been conducted by Calvin J. Heusser for the past three decades. His publication, in 1960, of *Late-Pleistocene Environments of North Pacific North America* has served as a framework on which to construct the post-Great Ice Age history of southern Alaska.

Beginning in 1957 as part of the International Geophysical Year, and continuing through the 1970s as part of the International Hydrological Decade, studies were conducted on McCall Glacier by a variety of institutions. McCall Glacier, in the Romanzof Mountains of the Brooks Range, is the northernmost investigated glacier in Alaska. In the Alaska Range, Gulkana Glacier was studied by numerous students and faculty from the University of Alaska under the direction of Troy L. Pewe from 1960 to 1964. From 1965 to present, Larry Mayo has studied Gulkana Glacier on a year-round basis.

A research team led by Dr. Mark Meier of the U.S. Geological Survey has recently invested much effort and money to determine how Columbia Glacier, largest in Prince William Sound, will behave in future decades. An increase in the production of icebergs from Columbia Glacier could create a severe navigational hazard to tankers transporting oil south from the port of Valdez. Captain Ronald Kolemeyer of the U.S. Coast Guard has examined the possibility of containing these icebergs by the use of a giant rope more than three miles long, stretched across the bay from the mainland to Glacier Island.

Dr. Carl Benson of the University of Alaska and others are studying the glaciers of Mount Wrangell and Redoubt Volcano to determine the relationship of glaciers to intermittently active volcanoes. Glaciers are used in these studies as "natural calorimeters" or heat flow sensors on the volcanoes.

In 1966 Larry Mayo began studies which continue today at Wolverine Glacier in the Kenai Mountains to determine the climate balance and river flow from the coastal maritime glacier. In 1973 Mayo and Dennis Trabant, also of the U.S. Geological Survey, initiated glacier motion studies at Gulkana, Wolverine, and Black Rapids glaciers to determine how glacier motion reacts to changes in climate.

1413 — CREVASSE IN MUIR GLACIER, ALASKA. FROM ABOVE.

MUIR GLACIER ON GLACIER BAY, ALASKA

BARTLETT GLACIER, ALASKA

12229. Child's Glacier, C. R. & N. W. Ry., Near Cordova, Alaska.
Published by O. Kennedy, Cordova, Alaska. Copyright, 1910, by O. Kennedy

Glaciers: A Magnet for Early Day Tourists

Editor's note: *Alaska's glaciers have attracted tourists for nearly a century, and postcards of those glaciers have permitted the visitors to show friends and family what they were missing. Here are some examples of those early day postcards, from the collection of Nicki Nielsen, with the thoughts of two tourists who sent them.*

Dear Elizabeth,
We arrived safely Mon. noon. Had a very stormy trip. Both seasick.

So begins the message on a postcard, bearing a photograph of several people in a small boat sailing before the face of Childs Glacier. Signed simply "Alice," the card is postmarked Cordova, Alaska, November 24, 1910. Alice continues:

Ground covered with snow when we came but fair now. Warm and rainy up to today. Clear and bright. Hope you are well. With love, Alice.

In 1910, tourism in Alaska was a mere shadow of the multi-million dollar industry it is today. But even then, visitors were drawn to the state's glaciers. They were exciting to visit, and postcards (usually photographs — some of which were hand-

colored) showing the gigantic masses of ice could share some of that excitement with the folks back home.

A later card, showing the spectacular terminus of Mendenhall Glacier, conveys an enthusiasm still felt by those traveling to Alaska today:

Tues. A.M., July 18, 1939
Dear Pauline,
You'd love this trip as it is so very interesting and different from any other place I ever visited. This A.M. I bought your bone faced doll with a fur coat — kinda homely. Hope your trip will be as interesting. Harriet.

Life on Alaska's Glaciers

Glaciers are not ecological deserts. Plants and animals live year-round on the surface of many Alaska glaciers.

Glacier worms or ice worms live on many glaciers of southeastern and south-central Alaska. The worm is only about one-half inch in length and must hatch its eggs in subfreezing temperatures. The worm's diet centers around algae living on the glaciers or other organic debris that is blown onto the ice surface.

Another animal inhabitant of Alaska's glaciers is the snow flea or glacier flea, a vegeterian insect that belongs to the order *Collembola*. The main ingredient of this insect's diet is conifer pollen, sometimes supplemented with red algae.

Red algae is actually a red-pigmented green algae able to make its own food through photosynthesis. During long, sun-filled days of summer, this snow algae can reproduce so rapidly that large areas are covered by a red snow condition. Meltwater streams draining red snowfields sometimes carry so much red algae that the water has a pinkish tinge. Such was the case on Cathedral Glacier, a small valley glacier that drains the eastern Juneau Icefield, in 1976.

Lichens play a very important role in determining the sequence of events in a glaciated area even though they do not grow on the ice. Lichens are compound plants made up of algae and fungus in a symbiotic relationship and are usually

Clockwise from left — *One of the first plants to become established in deglaciated areas is fireweed. This cluster was found growing in a recently deglaciated section of Muir Inlet.* ►*Lichens play an important role in determining the sequence of events in a glaciated area. By measuring this lichen* (Rhizocarpon geographicum), *which grows at a constant rate, the age of the lichen and the associated moraine can be established. With the aid of this information, it is possible to reconstruct the history of the glacier.* (Ransom Saltmarch) ►*This clump of moss, six inches in diameter, was found growing on bare bedrock, alongside Reid Glacier, in Glacier Bay National Park and Preserve.* ►*Bergs calved from tidewater glaciers provide safety from larger predators and secure pupping areas for harbor seals. These animals were observed on the ice near the face of Columbia Glacier in Prince William Sound.* (Tom Walker)

Far left — *Black-legged kittiwakes and glaucous-winged gulls, a common sight in Prince William Sound, perch on an iceberg calved from Shoup Glacier.* (Janet Ann Fries)
Left — *Ross avens is the only sign of life in this view of rock and snow surrounding Raven Glacier, just 27 miles from Anchorage.* (Jon R. Nickles)

found growing on bare rock surfaces. Studies by Roland Beschel in the 1950s showed that lichens were one of the first organisms to become established following ice retreat and that by measuring diameters of the largest members of known species, a growth curve could be constructed for a given geographic area. Dating of glacial events by analysis of lichen growth rates, the science of lichenometry, is one of the most effective relative age dating techniques used today, especially in areas where no radiocarbon datable trees are present. The three species most commonly used for analysis in Alaska are *Rhizocarpon, Umbilicarbius,* and *Lecidea.*

Alaska's Ice Worms

By Harvey Bowers, photo by the author

Most people outside the scientific community believe ice worms are a myth. Even some scientists question their existence; one microbiologist I know refused to believe in the worms until shown specimens.

While I spend a great deal of time around glaciers and snowfields, I never saw an ice worm until August, 1975, when I was coming down Byron Glacier late one evening. Black patches on the snow caught my attention. When I looked closely, the patches appeared to move. I pulled the flashlight from my pack and directed the light on hundreds of worms burrowing into the snow. In the light I could see more worms coming up through the snow. The largest of the worms was about the size of a spruce needle.

Adult ice worms are usually black, from three-quarters to one inch long, and about the diameter of a darning needle. Extremely sensitive to heat, the worms wriggle vigorously when caught in a handful of snow. Temperatures much below freezing seem to kill the worms. If kept in water just above freezing, ice worms will live for months without care. I have kept the tiny animals alive in a film canister half-full of water and refrigerated for eight months.

Ice worms in Alaska belong to the species *Oligochaete*, or earthworms, and are the only earthworms known to inhabit snow and ice. Specimens collected on Mount Rainier in Washington state are thought to be a new species *Anchytraeid*, or ice worm.

Scientists assume ice worms feed off algae growing on the snow, since the worms are normally found in areas of red algae. Several species of birds prey on the worms, picking them off the snow.

Iceworms, the only earthworms known to inhabit snow and ice, thrive in temperatures just above freezing. Adult worms are usually black and reach from three-quarters to one inch in length.

Members of a Canadian expedition to Greenland in 1872 first noted the ice worms. First Alaska record for this species was recorded by Dr. G.F. Wright on Muir Glacier in 1887. In 1891 Dr. I.C. Russell of the U.S. Geological Survey observed ice worms on an expedition to Malaspina Glacier. Russell reported, "In the early morning before the sunlight touched the snow its surface was literally covered with small, slim black worms, about an inch long, and having a remarkable snake-like appearance. These creatures were wiggling over the snow in thousands, but as soon as the sun rose and made its warmth felt, they disappeared beneath the surface."

On an 1899 expedition to Malaspina Glacier, Henry Bryant noted, "During the month of June and the early part of July, while the snow is comparatively dry, they appear about four o'clock in the afternoon on the surface and move sluggishly about, their dark color being quite conspicuous against the white background. They remain on the surface during the night; but when the sun appears in the morning they again burrow in the snow. They are widely but not uniformly distributed over the entire snowfield of the glacier."

Chances of seeing ice worms improve considerably on late evening trips to a glacier, but their tiny trails can also be seen in muddy surfaces such as meltwater pools. Remember to look the next time you drink from a meltwater pool, and watch where you step.

Alaska's Glaciers: A Closer Look

Coast Mountains

Most glaciers of southeastern Alaska, the area from the Canadian border at the south to the Fairweather Range at the north, are located in two massive ice fields of the Coast Mountains — the Stikine and Juneau icefields. Combined, these two ice fields contain more than 2,500 square miles of glacier-covered area within Alaska. Both ice fields also have substantial glacier-covered areas in British Columbia, Canada.

A number of smaller ice fields straddle the crest of the Coast Mountains to the north and southeast of the Stikine Icefield. Chickamin and Wright are the largest of Alaska's glaciers in these other small ice fields. Chickamin Glacier, located northwest of Hyder, has a length of more than 12 miles, eight of which are in Alaska, and an area of more than 50 square miles. At present, Chickamin is retreating and has developed a lake behind its terminal moraine. Field reports indicate that the glacier retreated almost 9,000 feet between 1902 and 1964, an average annual retreat of nearly 145 feet. Wright Glacier, south of the

Previous page — *Immense Bagley Icefield, in the eastern Chugach Mountains, occupies a linear trench from north of the Icy Bay-Mount Saint Elias region to southwest of the Tana River. Mount Saint Elias (18,008 feet) towers in the distance, at the center of the photo.* (Gil Mull)
Left — *Many small glaciers can be seen on the peaks above Witches Cauldron, near the Alaska-Canada border. Austin Post named the valley in 1965 because he found it to be "a sinister unworldly place."* (Jeffrey S. Johnson)

Taku River, covers an area of about 40 square miles. Since first observed in 1891, Wright Glacier has retreated more than two miles. However, the glacier's terminus position has remained stable since 1965.

Glaciers also exist on four islands of the Alexander Archipelago: Baranof, Kupreanof, Chichagof, and Admiralty. Baranof Island has the most glaciers and the largest glaciated area with more than 60 miles of its crest covered by small glaciers. None of the glaciers on any of the four islands have been studied in detail and, not surprisingly, none are named.

The Stikine Icefield

Spanning the crest of the Coast Mountains for about 120 miles from the Stikine River to the Whiting River, the Stikine Icefield contains more than a dozen glaciers with lengths of 10 miles or more. Many of the larger glaciers descend from elevations of more than a mile to sea level, where they terminate in the fiords of Frederick Sound or Stephens Passage.

LeConte Glacier in LeConte Bay, the southernmost glacier reaching Frederick Sound and also the southernmost active tidewater glacier in North America, is a prolific iceberg producer. The glacier is popular with tourists and is the most visited of the Stikine Icefield glaciers because of its proximity to Wrangell and Petersburg. LeConte Glacier has retreated about two and one-half miles since its position was first charted in 1887, but has stabilized in the past two decades. John Muir visited the glacier in 1879 and described it as one of the "most imposing" and "first class glaciers" he had ever seen.

Baird and Patterson glaciers both drain into Thomas Bay, 14 miles north of Petersburg. At the end of the 19th century Patterson Glacier

Four islands of the Alexander Archipelago — Baranof, Chichagof, Admiralty and Kupreanof — possess glaciers. The highest concentration is found along 60 miles of the crest of Baranof Island's mountains, shown here at their southern end. (Ernest Manewal)

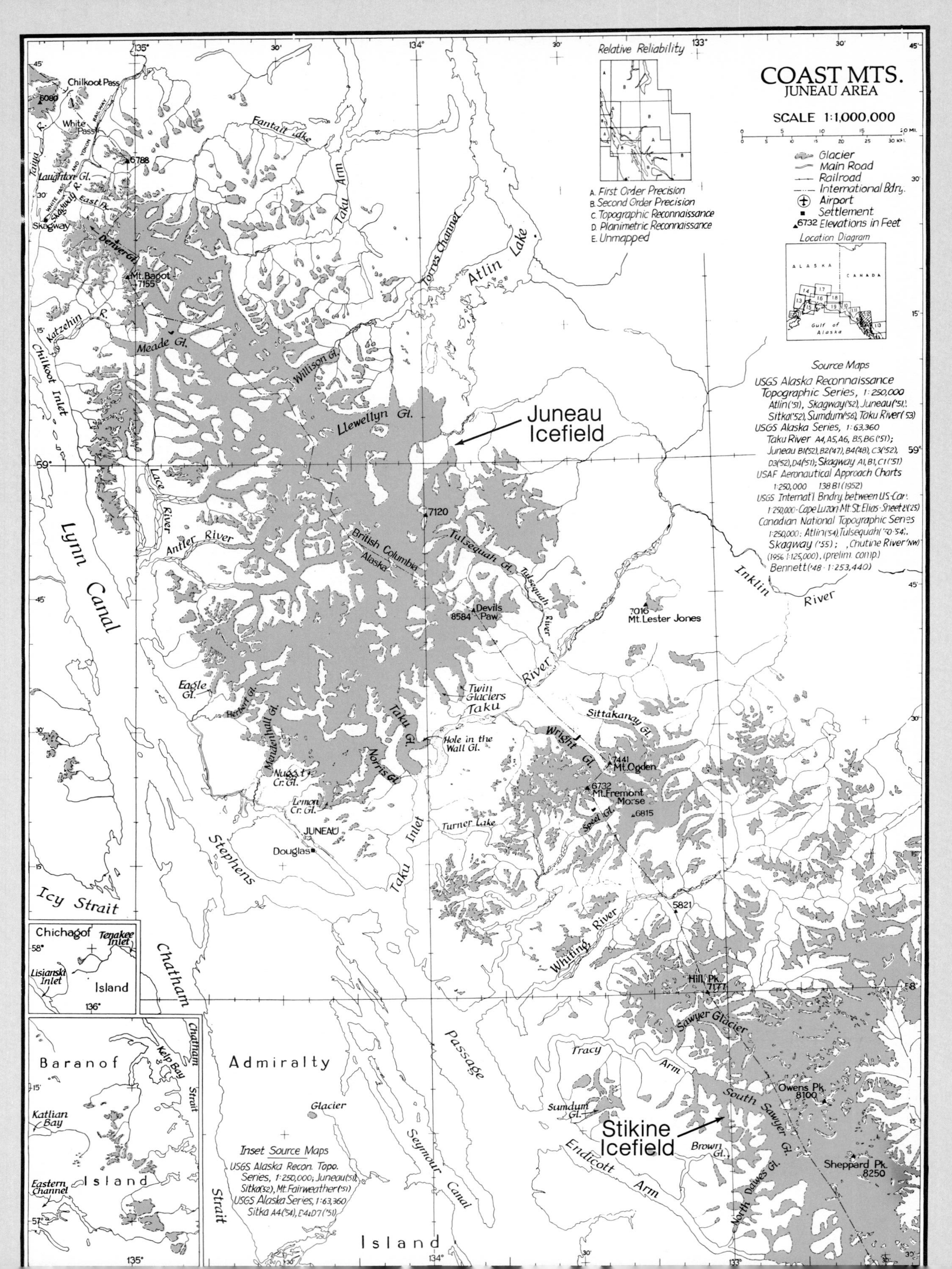

COAST MTS.
JUNEAU AREA
SCALE 1:1,000,000
Glacier
Main Road
Railroad
International Bdry.
Airport
Settlement
6732 Elevations in Feet
Location Diagram
ALASKA
CANADA
Gulf of Alaska
Relative Reliability
A. First Order Precision
B. Second Order Precision
C. Topographic Reconnaissance
D. Planimetric Reconnaissance
E. Unmapped
Source Maps
USGS Alaska Reconnaissance Topographic Series, 1:250,000 Atlin('51), Skagway('52), Juneau('51), Sitka('52), Sumdum('56), Taku River('53)
USGS Alaska Series, 1:63,360 Taku River A4, A5, A6, B5, B6 ('51); Juneau B1('52), B2('47), B4('48), C3('52), D3('52), D4('51); Skagway A1, B1, C1('51)
USAF Aeronautical Approach Charts 1:250,000 138 B1 (1952)
USGS Internat'l Bndry. between US-Can. 1:250,000 - Cape Luzon Mt. St. Elias - Sheet ('25)
Canadian National Topographic Series 1:250,000: Atlin('54), Tulsequah('50-'54), Skagway ('55); , Chutine River (NW) (1956 1:125,000), (prelim. comp.)
Bennett ('48 · 1:253,440)
Juneau Icefield
Stikine Icefield
Chilkoot Pass
6080
White Pass
Taiya R.
Laughton Gl.
Skagway R.
East Fk.
Skagway
6788
Denver Gl.
Mt. Bagot 7155
Katzehin R.
Meade Gl.
Chilkoot Inlet
Fantail Lake
Taku Arm
Torres Channel
Atlin Lake
Willison Gl.
Llewellyn Gl.
Lace River
Antler River
Lynn Canal
7120
British Columbia
Alaska
Tulsequah Gl.
Tulsequah River
Devils Paw 8584
7016 Mt. Lester Jones
Inklin River
Eagle Gl.
Herbert Gl.
Mendenhall Gl.
Twin Glaciers
Taku River
Taku Gl.
Hole in the Wall Gl.
Norris Gl.
Nugget Cr. Gl.
Lemon Cr. Gl.
JUNEAU
Douglas
Stephens
Taku Inlet
Turner Lake
Sittakanay Gl.
Wright Gl.
7441 Mt. Ogden
6732 Mt. Fremont Morse
6815
Speel Gl.
Icy Strait
5821
Whiting River
Chichagof
Tenakee Inlet
Lisianski Inlet
Island
Chatham
Hill Pk. 7171
Sawyer Glacier
Tracy Arm
Passage
Admiralty
Glacier
Island
Baranof
Kelp Bay
Chatham Strait
Katlian Bay
Eastern Channel
Island
Inset Source Maps
USGS Alaska Recon. Topo. Series, 1:250,000; Juneau('51), Sitka('52), Mt. Fairweather('51)
USGS Alaska Series, 1:63,360 Sitka A4('54), D4 D7('51)
Strait
Seymour Canal
Sumdum Gl.
South Sawyer Gl.
Owens Pk. 8100
Brown Gl.
Endicott Arm
North Dawes Gl.
Sheppard Pk. 8250

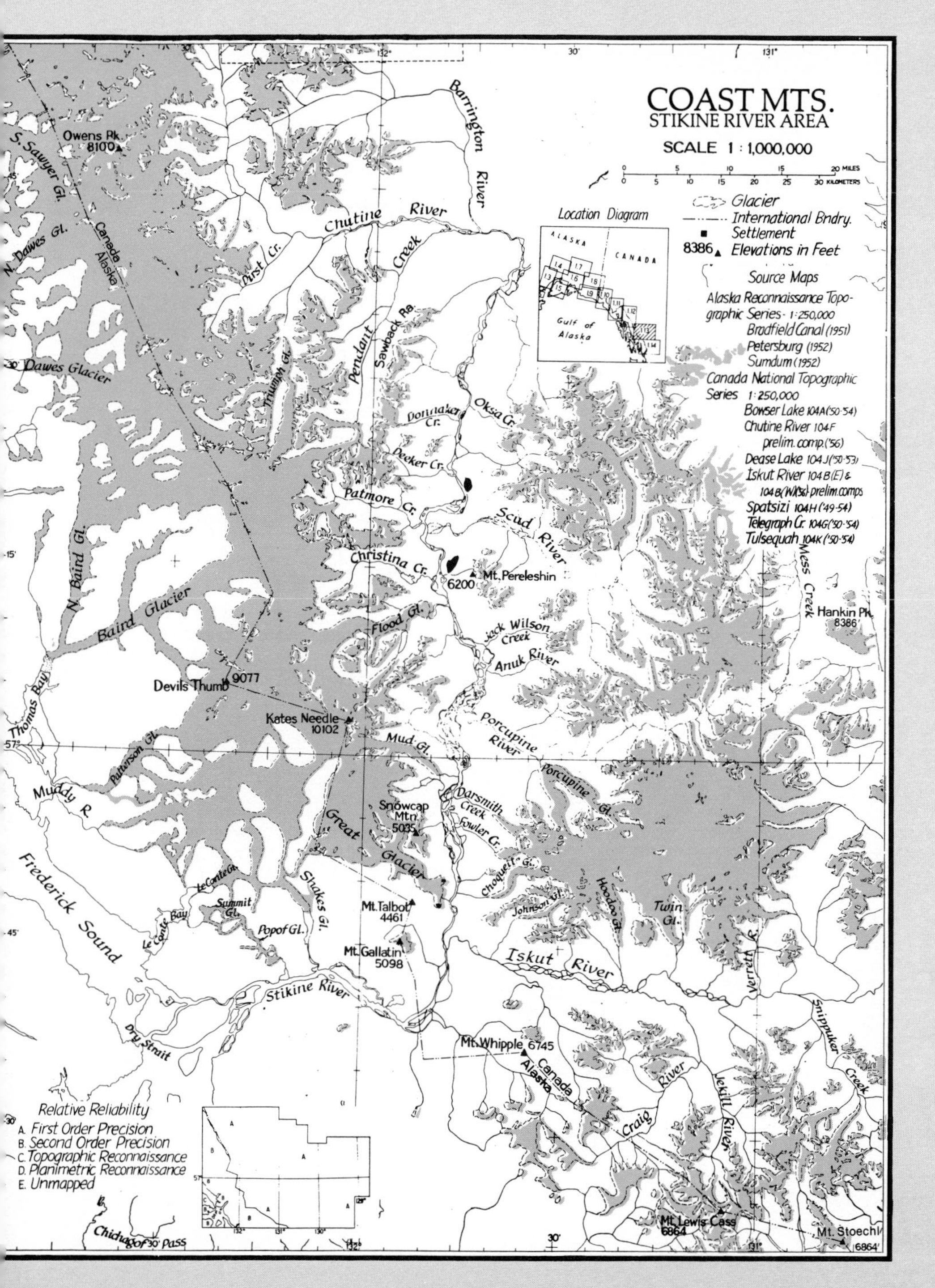

COAST MTS.
STIKINE RIVER AREA
SCALE 1 : 1,000,000
20 MILES
30 KILOMETERS
Glacier
International Bndry.
Settlement
8386 Elevations in Feet
Location Diagram
ALASKA
CANADA
Gulf of Alaska
Source Maps
Alaska Reconnaissance Topographic Series - 1:250,000
Bradfield Canal (1951)
Petersburg (1952)
Sumdum (1952)
Canada National Topographic Series 1:250,000
Bowser Lake 104A('50-'54)
Chutine River 104F prelim. comp.('56)
Dease Lake 104J('50-'53)
Iskut River 104B(E) & 104B(W)(s) prelim.comps
Spatsizi 104H('49-'54)
Telegraph Cr. 104G('50-'54)
Tulsequah 104K('50-'54)
Relative Reliability
A. First Order Precision
B. Second Order Precision
C. Topographic Reconnaissance
D. Planimetric Reconnaissance
E. Unmapped
Owens Pk. 8100
S. Sawyer Gl.
N. Dawes Gl.
Dawes Glacier
Canada
Alaska
Chutine River
First Cr.
Creek
Barrington River
Triumph Gl.
Pendant
Sawback Ra.
Oksa Cr.
Deeker Cr.
Patmore Cr.
Scud River
Christina Cr.
Mt. Pereleshin
6200
N. Baird Gl.
Baird Glacier
Flood Gl.
Jack Wilson Creek
Anuk River
Mess Creek
Hankin Pk. 8386
Devils Thumb 9077
Kates Needle 10102
Thomas Bay
Patterson Gl.
Muddy R.
Mud Gl.
Porcupine River
Porcupine Gl.
Snowcap Mtn. 5035
Darsmith Creek
Fowler Cr.
Great Glacier
Choquette Gl.
Hoodoo Gl.
Twin Gl.
Johnson Gl.
Frederick Sound
Le Conte Gl.
Le Conte Bay
Summit Gl.
Popof Gl.
Shakes Gl.
Mt. Talbot 4461
Mt. Gallatin 5098
Iskut River
Verrett R.
Stikine River
Dry Strait
Mt. Whipple 6745
Canada
Alaska
Craig River
Jekill River
Snippaker Creek
Mt. Lewis Cass 6864
Mt. Stoeckl
Chichagof Pass

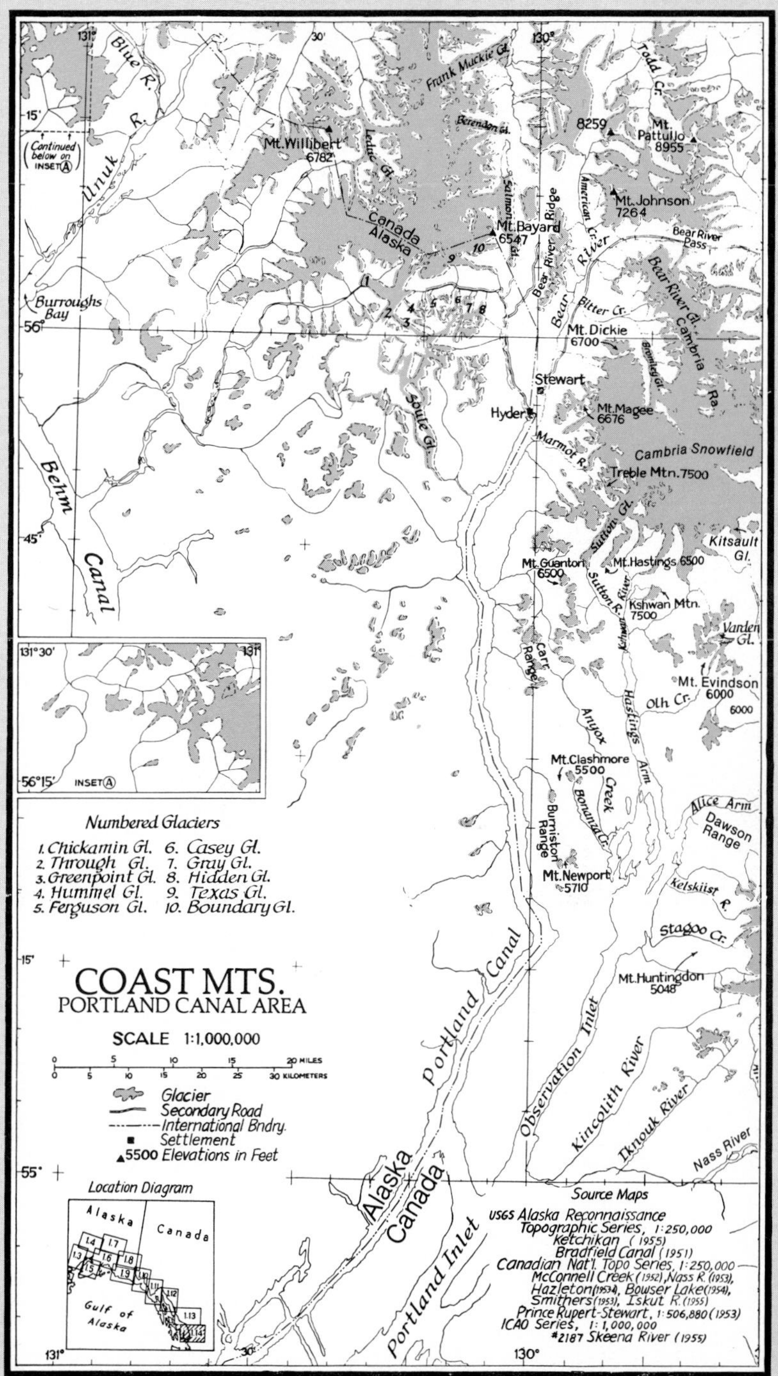

COAST MTS.
PORTLAND CANAL AREA
SCALE 1:1,000,000
20 MILES
30 KILOMETERS
Glacier
Secondary Road
International Bndry.
Settlement
5500 Elevations in Feet
Numbered Glaciers
1. Chickamin Gl.
2. Through Gl.
3. Greenpoint Gl.
4. Hummel Gl.
5. Ferguson Gl.
6. Casey Gl.
7. Gray Gl.
8. Hidden Gl.
9. Texas Gl.
10. Boundary Gl.
Location Diagram
Alaska
Canada
Gulf of Alaska
Source Maps
USGS Alaska Reconnaissance Topographic Series, 1:250,000
Ketchikan (1955)
Bradfield Canal (1951)
Canadian Nat'l. Topo Series, 1:250,000
McConnell Creek (1952), Nass R. (1953),
Hazleton (1954), Bowser Lake (1954),
Smithers (1953), Iskut R. (1955)
Prince Rupert-Stewart, 1:506,880 (1953)
ICAO Series, 1:1,000,000
#2187 Skeena River (1955)
Continued below on INSET A
INSET A
Blue R.
Unuk R.
Frank Mackie Gl.
Berendon Gl.
Mt. Willibert 6782
Leduc Gl.
Salmon Gl.
Canada
Alaska
Mt. Bayard 6547
Todd Cr.
8259
Mt. Pattullo 8955
Mt. Johnson 7264
American Cr.
Bear River Ridge
Bear River
Bear River Pass
Bear River Gl.
Bitter Cr.
Mt. Dickie 6700
Cambria Ra.
Burroughs Bay
Soule Gl.
Stewart
Hyder
Mt. Magee 6676
Marmot R.
Cambria Snowfield
Treble Mtn. 7500
Behm Canal
Kitsault Gl.
Mt. Guanton 6500
Mt. Hastings 6500
Sutton Gl.
Kshwan Mtn. 7500
Kshwan River
Vanden Gl.
Carr Range
Mt. Evindson 6000
Oth Cr.
Anyox
Hastings Arm
Mt. Clashmore 5500
Bonanza Cr.
Creek
Burniston Range
Alice Arm
Dawson Range
Mt. Newport 5710
Kelskiist R.
Stagoo Cr.
Mt. Huntingdon 5048
Portland Canal
Observation Inlet
Kincolith River
Iknouk River
Nass River
Alaska
Canada
Portland Inlet

Although the terminus of Great Glacier is on the Stikine River in British Columbia, the glacier heads in the Coast Mountains, in Alaska. Great Glacier is part of the Stikine Icefield, which spans the crest of the Coast Mountains for a distance of approximately 120 miles. (Staff)

LeConte Glacier is the southernmost active tidewater glacier in North America. Located in the Stikine Icefield, the glacier is a very popular tourist attraction due to its proximity to the towns of Petersburg and Wrangell. When John Muir visited LeConte Glacier in 1879, he described it as one of the "most imposing" glaciers he had seen. (Steve McCutcheon)

was advancing, but in the past 50 years it has retreated about one and one-quarter miles. Baird Glacier has a history of recent advance. A large outwash plain has been developing in front of Baird Glacier's terminus during this period of advance. To date, the length of the outwash plain is almost two miles. Prior to the 19th century, when the outwash plain's growth began, Baird was a tidewater glacier.

Two large glaciers lie at the head of Endicott Arm, a large southeast trending fiord that drains into Stephens Passage. The glaciers, Dawes and North Dawes, named after Henry Laurens Dawes, a politician from Massachusetts, have been studied since 1880 when they were first observed by John Muir. Dawes, a tidewater glacier, has been receding since the first mapping of its terminus. North Dawes was a calving tidewater glacier when it was first observed, but between 1889 and 1923 it retreated out of the water and has continued the recessional trend to the present time.

An example of rapid retreat is demonstrated by Brown Glacier, which had a tidewater terminus at the head of Fords Terror, a side fiord of Endicott Arm. In 1880 John Muir observed "shattered overleaning fragments" of ice falling from the tidewater terminus, but he did not map the glacier's terminal position. In 1894 the International Boundary Survey did record the position of the glacier's terminus. Fifteen years later they remapped the terminal position and discovered that Brown Glacier had retreated about one and one-half miles. Photographs taken by William O. Field in 1935 showed a total retreat of about three and one-half miles from the 1894 position. By 1950 retreat had exceeded four miles. Rapid retreat also characterizes South Brown Glacier, which separated from Brown Glacier between 1923 and 1929. South Brown Glacier receded more than a mile between 1929 and 1948.

Two large tidewater glaciers dominate the head of Tracy Arm, a long, sinuous fiord just north of Endicott Arm. The two glaciers, Sawyer and South Sawyer, now four miles apart, each have lengths of about 20 miles and areas of more than 100 square miles. The glaciers may have been joined within the past 200 years, possibly as recently as about 1880. Positions of both glaciers have fluctuated within the 20th century, but both have receded more than 25 miles from the time of their most recent maximum advance. In 1980 the positions of both Sawyer and South Sawyer glaciers were stable.

The Juneau Icefield

Located along the crest of the Coast Range from the Taku River to just east of Skagway, the Juneau Icefield is the most studied ice field in Alaska and among the most studied in the world. The ice mass includes the state's, and perhaps the country's, most visited glacier, Mendenhall. The ice field, which like the

Two Women and The Glacier

By Dr. Cyrus Peck, Sr.
and KTOO-FM, Juneau
Illustrations by Raymond E. Peck

Editor's note: *Alaska's Native people have had to deal with glaciers for many centuries. In one lecture from a 13-part Tlingit Culture Lecture Series produced by Cyrus Peck, Jr., for KTOO-FM, Juneau, Dr. Cyrus Peck, Sr., tells how his people overcame a glacier barrier many years ago.*

Clan members place a spruce tree decorated with feathers on the raft on which the two old women would ride down the river and under the glacier.

Swift runners from the clan wait on the glacier for two days to watch for the raft and two old women to emerge from under the glacier.

There was a great glacier across the Stikine River, and as they [the Tlingits] migrated down the Stikine River they came face to face with the great glacier, the face of the glacier, and the Stikine River flowing under the glacier. So they came to a stop. Their leader said halt all movement — any further down the river they go, it would be disaster for them to venture under the glacier. They wanted to go on. The younger generation wanted to move, but the much more sober, more sensible seniors of that particular clan said, "No, let's stop here." They made camp alongside of the Stikine River at the face of the glacier. No matter what they did and what plans they made, it just didn't seem right that the glacier was still a killer and they cannot go under it blind.

Finally, one day two little old women came to the clan leader saying, "We are much advanced in years now. We would like to take the honor of being placed on a raft, and to drift down under the glacier and have runners to see and wait on the other side of the glacier. If we come out alive, then the entire tribe can move and migrate under the glacier."

The leader said, "No." He had to make the decision a very, very stern one. It was either they move or they stop there. Finally he called a council . . . and they decided, "Well, they volunteered, we can't refuse. Bring them before the council." So they did. The two little old women were brought in and they faced the council, and they asked them questions. "Have you changed your mind, or could we change your mind from going under there? Where we can't stop you if you want to go, we can make you as comfortable as possible on the raft, but we can't force you." They said, "We have decided, for the future generation, that we go under the glacier."

After a unanimous vote, they were given permission. The villagers prepared the raft to go under the glacier, and they prepared these two women by meeting with them, singing the songs and dance, farewell songs and dance — the songs that they sing are songs of the creation of the emblem. Songs of creation of their identity, who they are.

When everything was over, they had the raft ready. And they placed a spruce tree, as you would a Christmas tree, in the middle of the raft, and they put feathers all over the tree. And when they were ready, they started the ceremony. They started singing and dancing, and when all this was over, they let the raft go. The two little women sitting on the raft had composed their own song for the occasion, and they had a drum, and they sang as the raft drifted down under the glacier and out of sight.

Two young men came forward and said, "We consider ourselves the fastest runners of this tribe, and we ask permission to go over the glacier, down to the other end and watch for them, so that we might bring word whether the raft comes out broken up and the women are not on the raft. Or if they are alive, we are going to bring the word back."

So the permission was given, and they ran and waited at the other end. Waited and waited. A day passed. Second day, toward evening, the raft came out with the two women sitting and singing and the spruce tree covered with feathers untouched. The runners went back and made the sound of victory as they went over and approached the village that these two women have made it under the glacier.

Hikers cross a meltwater stream of Herbert Glacier, north of Juneau, by means of glacial boulders. Studies indicate that the glacier has been retreating at the rate of 60 feet per year since 1766.

Stikine also extends into British Columbia, covers an area of more than 1,215 square miles in Alaska and has over 30 valley glaciers that descend to near sea level. Glaciers draining the Juneau Icefield are retreating, with the exception of Taku Glacier which is presently advancing across Taku Inlet at a rate of a few hundred feet per year.

Taku Glacier, more than 30 miles long, is the largest glacier flowing out of the Juneau Icefield. In late 1980 the glacier's terminus was steadily advancing across Taku Inlet at a rate that would completely close the inlet and create a large, glacier-dammed lake in the next three to four decades. As Taku advances, it is knocking down trees in the forests on both edges and is overriding the moraines and outwash plain of Norris Glacier, immediately to its south. Taku Glacier advanced across Taku Inlet in a similar manner about 200 years ago, forming a large lake as evidenced by shoreline high on the sides of the inlet. The vegetative history and the history of the formation of the lake has been investigated by Donald B. Lawrence of the University of Minnesota.

Taku Glacier's present advance began about 1890. At that time, and until the mid-1930s, Taku Glacier ended in a deep fiord and calved spectacular quantities of icebergs into Taku Inlet. But sediment produced by the advancing glacier began filling the inlet to the point that, by 1939, cruise ships, which previously had sailed right up to the glacier's terminus, could

not enter the inlet. Since the 1950s Taku's advancing terminus has sat upon a **push moraine** that has prevented icebergs from calving off the terminus. This reduction in ice loss from the glacier is an important factor in the ongoing advance across Taku Inlet. John Muir said: "To see this one glacier is well worth a trip to Alaska." Today, for different reasons, this is still the case.

Hole-in-the-Wall Glacier, so named because it formed when an arm of Taku Glacier thickened and expanded over the valley wall in which it had previously been contained, is a distributary lobe of Taku Glacier. An extremely short Hole-in-the-Wall Glacier may have existed in 1890. The present glacier overtopped its valley wall in about 1940, and since then has continued to advance to the level of the Taku River.

The presence of Taku and Hole-in-the-Wall glaciers has created insurmountable problems for construction of a road to connect Juneau, largest community in southeastern Alaska and the state's capital, to the Alaska Highway via the Taku River Valley. Consequently, Juneau is the only state capital in the country not accessible by road.

Norris Glacier, adjacent to Taku, has a totally different advance and retreat history than its larger neighbor. Norris stood at its

Hole-in-the-Wall Glacier, a distributary of Taku Glacier, overtopped its valley wall around 1940 and has continued to advance since then. The presence of these two glaciers has been the main obstacle in the construction of a road through the Taku River valley to connect Juneau with the Alaska Highway. Consequently, Juneau is the only state capital in the United States not accessible by road.

most recent maximum advance position about 1917. Since that time, Norris has retreated about one mile and has thinned appreciably. A large branch of Norris, called the Dead Branch, was once an active, ice-supplying tributary to the main Norris Glacier. In late 1980 the branch was an essentially stagnant distributary arm that received ice from the main glacier. Trimlines suggest that this reverse ice flow has taken place for at least the last 100 years.

Although side by side, Taku and Norris are significantly different. Taku has a much larger and much higher area of accumulation than Norris. In fact, 40% of Taku's accumulation area is above 4,500 feet, while only 2% of Norris' is that high.

Lemon Creek Glacier, one of the better studied small glaciers in the Juneau Icefield, has a history of detailed observations extending back to 1953. The glacier, which can be seen high above Gastineau Channel at the head of Lemon Creek Valley, has been retreating for about the last 230 years, with a total recession of more than one and one-half miles during that time.

Herbert and Eagle glaciers, north of Juneau, have histories similar to that of Lemon Creek Glacier. Studies by David B. Lawrence between 1949 and 1952 showed that Herbert retreated an average of 60 feet a year since 1766, with a maximum average rate of retreat of 190 feet per year between 1928 and 1948.

Less than 15 miles from downtown Juneau is Mendenhall Glacier, named for a former

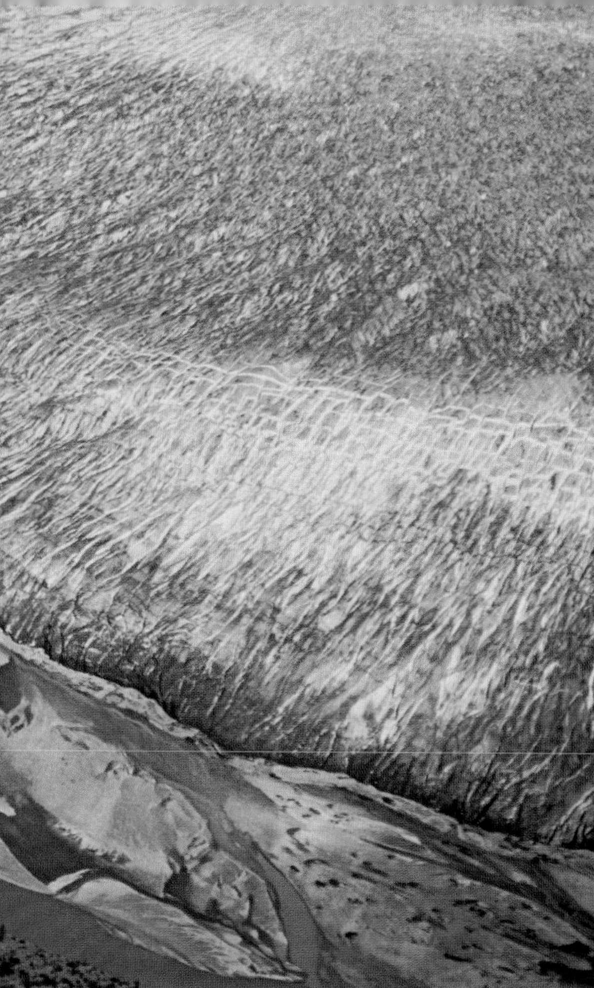

Left — *Although the majority of glaciers draining the Juneau Icefield are retreating, Taku Glacier is advancing at a rate of a few hundred feet per year. If it continues to move forward at this rate, it could completely close Taku Inlet within the next 30 or 40 years, creating a large, glacier-dammed lake.*

Below — *Two small, unnamed glaciers flow from the summit of Mount Foster (7,127 feet), top right, located on the Alaska-Canada border, about 25 miles north of Skagway.* (Wesley K. Wallace)

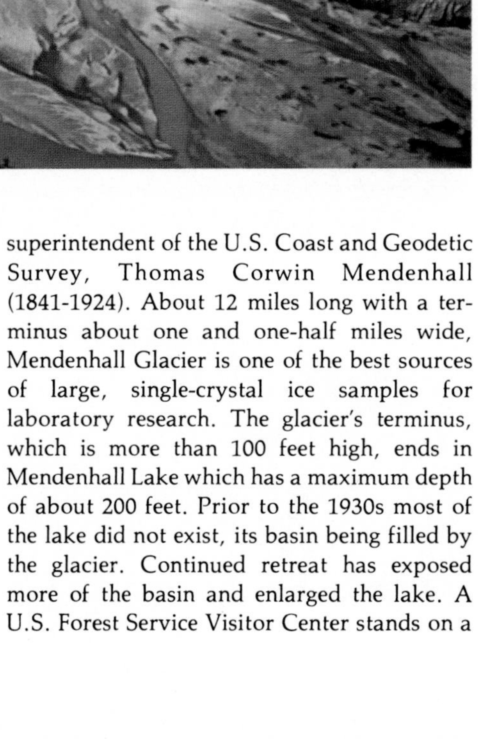

superintendent of the U.S. Coast and Geodetic Survey, Thomas Corwin Mendenhall (1841-1924). About 12 miles long with a terminus about one and one-half miles wide, Mendenhall Glacier is one of the best sources of large, single-crystal ice samples for laboratory research. The glacier's terminus, which is more than 100 feet high, ends in Mendenhall Lake which has a maximum depth of about 200 feet. Prior to the 1930s most of the lake did not exist, its basin being filled by the glacier. Continued retreat has exposed more of the basin and enlarged the lake. A U.S. Forest Service Visitor Center stands on a spot that was covered by ice as recently as 1940. Since then Mendenhall has retreated about one-half mile and is now separated from the visitor center by the lake. Today Mendenhall is retreating at a rate of less than 100 feet per year and has ice flowing through it at a rate of less than five feet per day. Work by Lawrence published in 1950 recognized 13 distinct recessional moraines in the slightly more than two miles between Mendenhall's 18th century terminal moraine position and the late 1940s ice position.

The terminus of Mendenhall Glacier, which is about 100 feet high, ends in Mendenhall Lake. This photo was taken from near the U.S. Forest Service Visitor Center which stands on a spot that was covered by ice as recently as 1940. In addition to the visitor center, the forest service maintains several nature trails for those who wish to get a closer look at the glacier. (Kent Krone)

The Juneau Icefield Research Program

The Juneau Icefield Research Program (JIRP) was organized and implemented in the late 1940s to pursue detailed long-term research on the glacier systems of the Juneau Icefield. Primary emphasis has been on studies of the mechanics of glacier formation and movement, climatology, regional geology, ice field biology and botany, polar medicine, and many other topics. The program was originally established by the American Geographical Society as the Juneau Icefield Research Project and received financial and logistical support from the U.S. Navy, U.S. Army, and U.S. Air Force. In recent years, much support has been received from the National Science Foundation, the National Geographic Society, and the Foundation for Glacier and Environmental Research.

Field investigations began in 1948 and have continued to the present. The Summer Institute of Glaciological and Arctic Sciences was organized in 1959 to provide combined academic and field training for college and university students. This concept was expanded in the 1970s to include gifted high school students. The program is affiliated with the University of Idaho and offers both graduate and undergraduate credit.

Many of the leading professional scientists in the fields of glaciology, glacial geology, and arctic sciences have received their first glacial field experience and other important parts of their training through the Juneau Icefield Research Program. The program is under the expert direction and leadership of Maynard M. Miller, Dean of the College of Mines at the University of Idaho.

Camp 17, one of the Juneau Icefield Research Program field stations, is located on a small rock knob overlooking Lemon Creek and Ptarmigan glaciers. The program was organized in the late 1940s to carry out long-term research on the ice field's glacier systems. (Mark Kelley)

Counterclockwise from left — *Snow pits are dug to determine the thickness of snow that fell the previous winter and to inspect the firnification process.* ►*Cores of snow are taken at set intervals to detect changes in the density of the snow.* ►*Each core is carefully weighed and the density of the firn is calculated.* ►*When snow pits cannot be dug, ice and snow cores are collected and inspected.* ►*To determine the thickness of a glacier, hydrophones are placed on its surface and an explosive charge is detonated. The ice thickness can be calculated by measuring the time difference between the different types of sound waves. Here, the hydrophones are being deployed as a whiteout sets in.* ►*A group of geologists examine a cavity under the terminus of Gilkey Glacier, where they can observe the bedrock surface being abraded and polished.* ►*In the snowfield and* neve *areas, the best means of travel is by over-snow vehicle.*

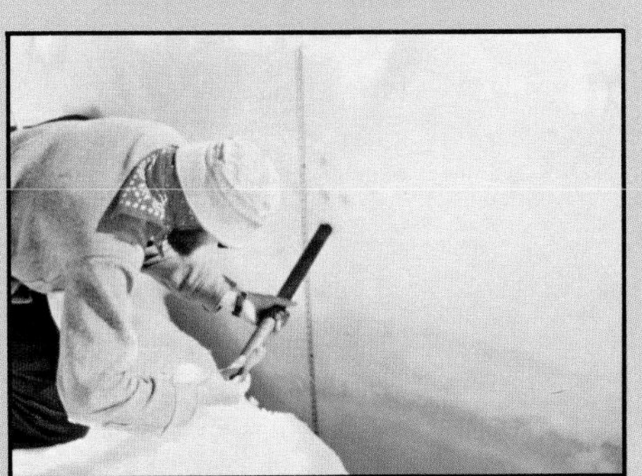

Saint Elias and Wrangell Mountains

The Saint Elias Mountains form a 300-mile-long and 90-mile-wide barrier that extends northwest along the Alaska-Canada boundary from Cross Sound, Icy Strait, and Lynn Canal on the southeast to Bering Glacier, Tana Glacier, Tana River, Chitistone River, and White River on the northwest. Alaska has about 60% of the Saint Elias Mountains. Mount Logan (19,850 feet), the range's highest peak, lies in Canada, but lofty summits that straddle the international boundary or are in Alaska include Mount Saint Elias (18,008 feet), Mount Bona (16,421 feet), Mount Vancouver (15,700 feet), Mount Fairweather (15,300 feet), Mount Hubbard (14,950 feet), and more than two dozen other peaks with elevations greater than 10,000 feet. In Alaska, the Saint Elias Mountains divide into the Fairweather Range, Chilkat Range, Takhinsha Mountains, Brabazon Range, and Saint Elias Range.

In Alaska more than 5,500 square miles of glaciers including more than 50 glaciers with lengths greater than five miles flow from the Saint Elias Mountains. Some of the area's more magnificent glaciers are Hubbard, the gigantic Malaspina complex, the glaciers of Glacier Bay, Brady Glacier, La Perouse Glacier, and Guyot and Yahtse glaciers of Icy Bay.

Glacier Bay

Glacier Bay, perhaps the most spectacular fiord complex in the world, contains 13 active, calving, tidewater glaciers within its 65-mile length and has the best studied, most completely documented history of recent glacier fluctuations in the state. About 200 years ago Glacier Bay did not exist because a single, gigantic ice field extended into Icy Strait and completely filled the bay. By 1794, when Vancouver's lieutenants Whidbey and LeMesurier explored Icy Strait, the ice had retreated, probably about six miles to the vicinity of Rush Point, and had opened a small bay that is shown on Vancouver's map of the northwest coast. In 1879 when John Muir explored the bay, the ice retreat had exceeded 40 miles. By late 1980 the ice had retreated more than 60 miles, but not all of the glaciers are still retreating. Observations made in 1980 showed that Muir, Riggs, and McBride glaciers in Muir Inlet had retreated from their 1979 positions, while Johns Hopkins and Reid glaciers had advanced.

The rivers of ice in Glacier Bay head in both the Fairweather and Takhinsha mountains, with many glaciers having parts of their areas of accumulation at elevations above 5,000 feet. At least 10 glaciers within the bay have lengths greater than 10 miles. The longest, Grand Pacific Glacier, heads in Alaska, passes through British Columbia, and terminates in Alaska, in Tarr Inlet.

Fiords of Glacier Bay that have actively calving, tidewater glaciers are Muir, Tarr, Johns Hopkins, Reid, and Wachusett inlets.

Logan Glacier, which heads in Canada and terminates in Alaska, is presently retreating. The glacier's large, pitted ablation moraine can be seen here. (Gil Mull)

Rendu, Queen, Adams, Hugh Miller, and Geikie inlets all had tidewater glaciers during the early part of this century, but in all cases the ice masses have retreated onto land.

JOHNS HOPKINS INLET: It is hard to say whether Johns Hopkins Inlet or Muir Inlet is the most spectacular in Glacier Bay. In terms of number of glaciers, Johns Hopkins wins hands-down. In the 10 miles between Lamplugh Glacier at the inlet's mouth and Johns Hopkins Glacier at the inlet's head, nine separate ice masses descend the walls of the fiord. These glaciers range in size from small, hanging glaciers such as Charley, John, and Clark glaciers, to active, calving, tidewater glaciers such as Toyatte, Kashoto, Hoonah, and Gilman. In front of Gilman Glacier previous glacial erosion has scoured the fiord bottom to a depth of more than 1,200 feet below sea level.

At the turn of the century Johns Hopkins Inlet was completely filled by Johns Hopkins Glacier and its hanging valley tributary glaciers. Between 1892 and 1929 these glaciers retreated more than 11 miles. For the last half century, Johns Hopkins Glacier has been slowly advancing and thickening with more than two miles of advance occurring since 1935.

John, Charley, and Toyatte glaciers were named for various Indians who helped Muir in his explorations in Alaska in 1879. Gilman and Clark glaciers were named for the first president and a professor of geology at Johns Hopkins University respectively, and Hoonah Glacier was named for the Hoonah tribe of Tlingit Indians.

TARR INLET: At the head of Tarr Inlet are Margerie and Grand Pacific glaciers, both of which are actively calving, tidewater glaciers. The two were joined until about 1912 when retreating Grand Pacific Glacier separated from Margerie. Continued retreat of Grand Pacific brought its terminus north of the international boundary into Canada sometime between 1913 and 1916. Grand Pacific's terminus remained in Canada until about 1948 when it advanced to straddle the border. Since 1961 Grand Pacific has very slowly re-advanced to a position about one mile into Alaska. Today Grand Pacific and Margerie glaciers are within 300 feet of rejoining, following almost 70 years of separation. Grand Pacific remains unique in that no other tidewater glacier in Alaska leapfrogs across the international boundary in such a fashion.

REID INLET: Reid Inlet is about two miles long and contains one glacier, Reid, at its head. A large terminal moraine located at the mouth of the inlet, about two miles from Reid Glacier's present position, formed between 1899 and the 1920s when the ice front sat nearly stationary. In late 1980 Reid Glacier appeared to be slowly re-advancing.

WACHUSETT INLET: Wachusett Inlet, a side arm of Muir Inlet, was formed by very rapid disintegration and retreat of Plateau Glacier during the last 50 years. Since 1929 the inlet was increased in length by more than a dozen miles. A stagnating arm of Carroll Glacier is presently calving small quantities of ice at the head of Wachusett Inlet. If Carroll Glacier continues to retreat, Wachusett Inlet might connect to Queen Inlet, the fiord into which the main part of Carroll Glacier presently drains. For this to happen the large outwash sediment mass that fills the junction between the two fiords will have to be removed by submarine landslides and the erosive action of tidal currents and waves.

Fairweather Range

The Fairweather Range stretches along the Gulf of Alaska from Cross Sound to the Alsek River, and contains La Perouse Glacier, the only calving glacier in North America that discharges icebergs directly into the open Pacific Ocean. Flowing from the Fairweather's mountainous heights are Brady Glacier, largest in the range; Grand Plateau and Fairweather glaciers, small piedmont glaciers that almost reach the Pacific Ocean; and a

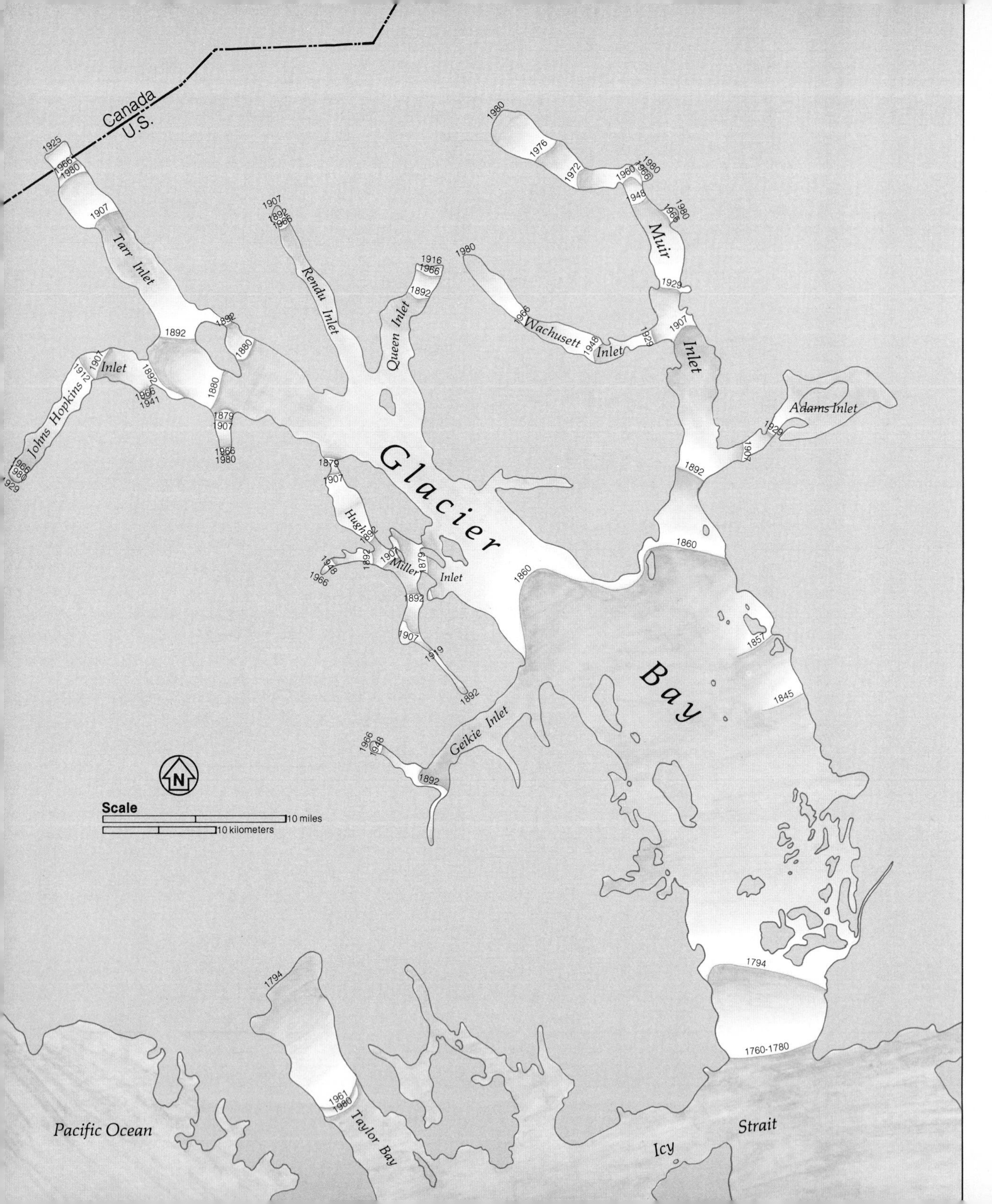

Figure 8 — *Ice positions in Glacier Bay from 1760 to 1980.*

number of glaciers that fill the trench of the Fairweather Fault, including the ice masses of Lituya Bay.

Hundreds of glaciers flow from the flanks of Fairweather Range peaks. Nowhere else on earth do glacier-covered mountains rise so abruptly, from sea level to heights of more than 10,000 feet, in some cases in distances of less than 10 miles. Within the Fairweather Range rise Mount Fairweather (15,300 feet), Mount Root (12,860 feet), Mount Crillon (12,726 feet), Mount Watson (12,516 feet), Mount Salisbury (12,170 feet), Lituya Mountain (11,924 feet), and Mount La Perouse (10,728 feet).

Brady Glacier, with an area of about 175 square miles, was a tidewater glacier at the time of Vancouver's exploration. During the 19th century Brady advanced more than four miles and developed a large outwash plain in front of its terminus. In late 1980 the glacier was slowly advancing, with huge quantities of sediment pouring out of its meltwater streams rapidly filling the upper end of Taylor Bay. Parts of Taylor Bay that had more than 300 feet of water in 1926 are now covered by tidal flats and outwash plain.

La Perouse Glacier heads on the flanks of Mount La Perouse and Mount Dagelet and descends to sea level where its three-and-one-half-mile-wide terminus is frequently washed by high tide and storm waves. In recent years the position of the ice terminus has fluctuated

With a tremendous splash, chunks of ice calve from the face of Muir Glacier. Two kayakers, insignificant by comparison, brace themselves for the wave that will follow. (Tom Bean)

Left — *A helicopter carrying a crew of geologists sets down in Deception Hills for a look at the broad, flat Grand Plateau Glacier. The peak of Mount Fairweather (15,300 feet) rises above the clouds in the distance.* (Gil Mull, reprinted from *ALASKA GEOGRAPHIC®*)
Right — *This aerial view shows Margerie Glacier, at the head of Tarr Inlet. In the bottom right corner of the photo is Grand Pacific Glacier, within 300 feet of joining Margerie. The two glaciers were joined until 1912, when Grand Pacific's retreat separated them.* (Tom Bean)

a few hundred yards, but has always ended at the surf zone. In 1966 La Perouse Glacier made a mini-surge and ended in an ice cliff about 300 feet high at the lowtide line.

Lituya Bay, a T-shaped fiord, contains three tidewater glaciers: Lituya, North Crillon, and Cascade. When La Perouse mapped the bay in 1786, a pair of separate glaciers terminated at or near tidewater in each of the upper ends of the T. In both instances the pair of glaciers have joined and advanced. Together North Crillon Glacier, in the east arm of the T, and Lituya Glacier, in the west arm, have advanced about six miles since 1786.

In 1958 a gigantic earthquake-generated rockslide fell in Gilbert Inlet, the west arm of Lituya Bay, and on the terminus of Lituya Glacier. The resulting splash wave removed soil and vegetation from surrounding slopes up to an elevation of more than 1,750 feet.

Saint Elias Range

The Saint Elias Range extends from Russell Fiord, the eastern extension of Yakutat Bay, to the Wrangell and Chugach mountains and includes Mount Saint Elias (18,008 feet), highest mountain in the Alaska part of the range, and Malaspina Glacier, largest piedmont lobe in North America. The range also encompasses two large, glacially eroded bays, Yakutat and Icy, and many surging glaciers including Walsh Glacier.

Walsh Glacier, located astride the Canadian border 60 miles north of Icy Bay, is about 35 miles long and covers an area of 160 square miles. From 1960 to 1966 Walsh Glacier underwent a surge that resulted in ice movement of about 10½ miles with a maximum advance of 2½ miles between August 1965 and September 1966. Although scientists have suggested that Walsh Glacier's surge was triggered by the 1964 earthquake, Austin Post has

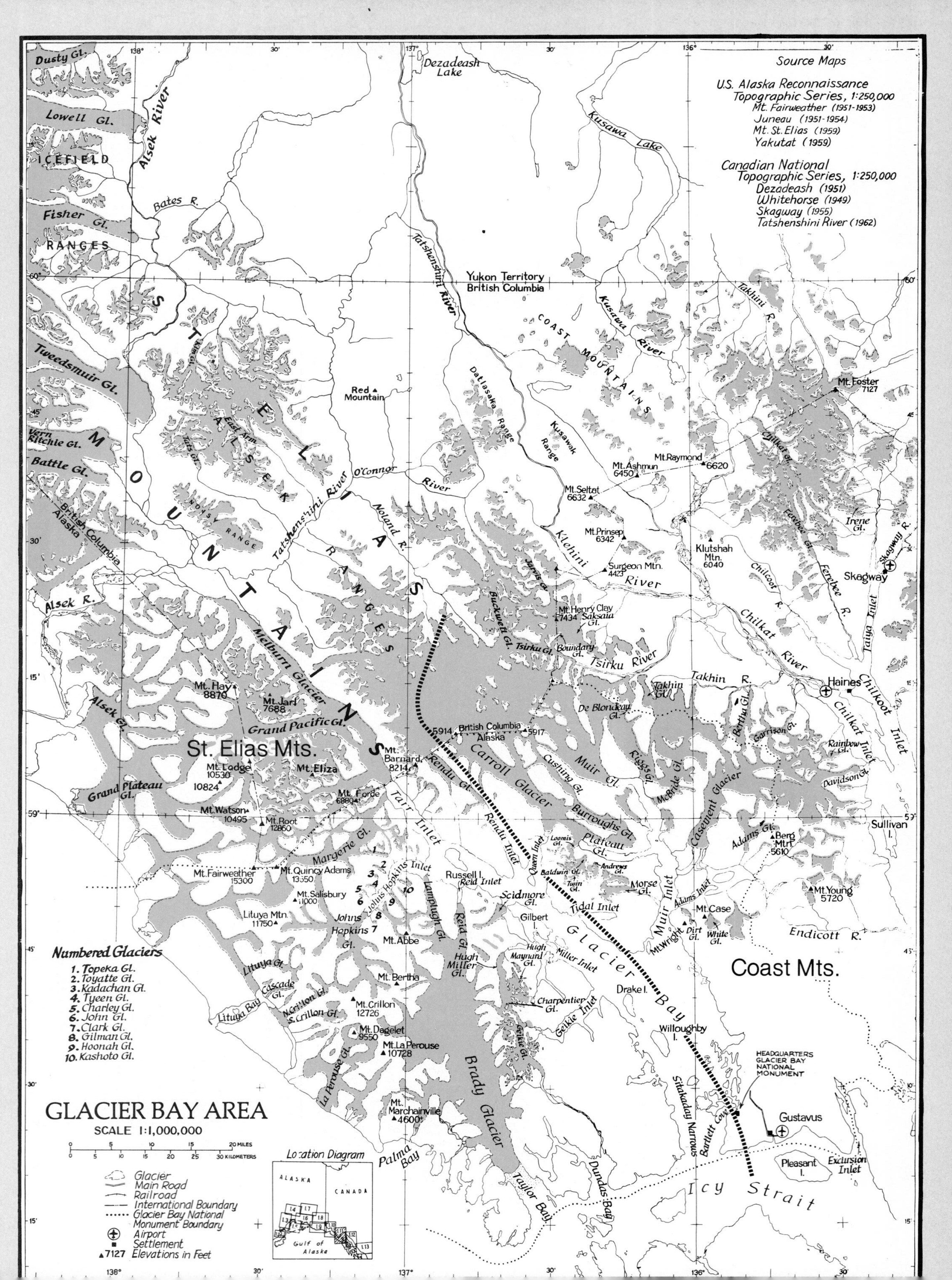
GLACIER BAY AREA
SCALE 1:1,000,000
0 5 10 15 20 MILES
0 5 10 15 20 25 30 KILOMETERS
Glacier
Main Road
Railroad
International Boundary
Glacier Bay National Monument Boundary
Airport
Settlement
7127 Elevations in Feet
Location Diagram
ALASKA
CANADA
Gulf of Alaska
Numbered Glaciers
1. Topeka Gl.
2. Toyatte Gl.
3. Kadachan Gl.
4. Tyeen Gl.
5. Charley Gl.
6. John Gl.
7. Clark Gl.
8. Gilman Gl.
9. Hoonah Gl.
10. Kashoto Gl.
Source Maps
U.S. Alaska Reconnaissance Topographic Series, 1:250,000
Mt. Fairweather (1951-1953)
Juneau (1951-1954)
Mt. St. Elias (1959)
Yakutat (1959)
Canadian National Topographic Series, 1:250,000
Dezadeash (1951)
Whitehorse (1949)
Skagway (1955)
Tatshenshini River (1962)
Coast Mts.
St. Elias Mts.
ST. ELIAS MOUNTAINS
COAST MOUNTAINS
ALSEK RANGES
ICEFIELD RANGES
Brady Glacier
Carroll Glacier
Casement Glacier
Melburn Glacier
Grand Pacific Gl.
Grand Plateau Gl.
Glacier Bay
Icy Strait
Muir Inlet
Tarr Inlet
Rendu Inlet
Queen Inlet
Tidal Inlet
Johns Hopkins Inlet
Dundas Bay
Taylor Bay
Palma Bay
Lituya Bay
Sitakaday Narrows
Bartlett Cove
Excursion Inlet
Chilkoot Inlet
Chilkat Inlet
Taiya Inlet
Haines
Skagway
Gustavus
HEADQUARTERS GLACIER BAY NATIONAL MONUMENT
Mt. Fairweather 15300
Mt. Quincy Adams 13560
Lituya Mtn. 11750
Mt. Salisbury 11000
Mt. Crillon 12726
Mt. Dagelet 9550
Mt. La Perouse 10728
Mt. Marchainville 4600
Mt. Root 12860
Mt. Watson 10495
Mt. Lodge 10530
Mt. Hay 8870
Mt. Jarl 7688
Mt. Forde 6880
Mt. Barnard 8214
Mt. Foster 7127
Mt. Raymond 6620
Mt. Ashmun 6450
Mt. Seltat 6632
Mt. Prinsep 6342
Surgeon Mtn. 4423
Klutshah Mtn. 6040
Mt. Henry Clay 7434
Berg Mtn. 5610
Mt. Young 5720
Red Mountain
Yukon Territory
British Columbia
Alaska
Tatshenshini River
Alsek River
Klehini River
Tsirku River
Chilkat River
Dezadeash Lake
Kusawa Lake
Endicott R.

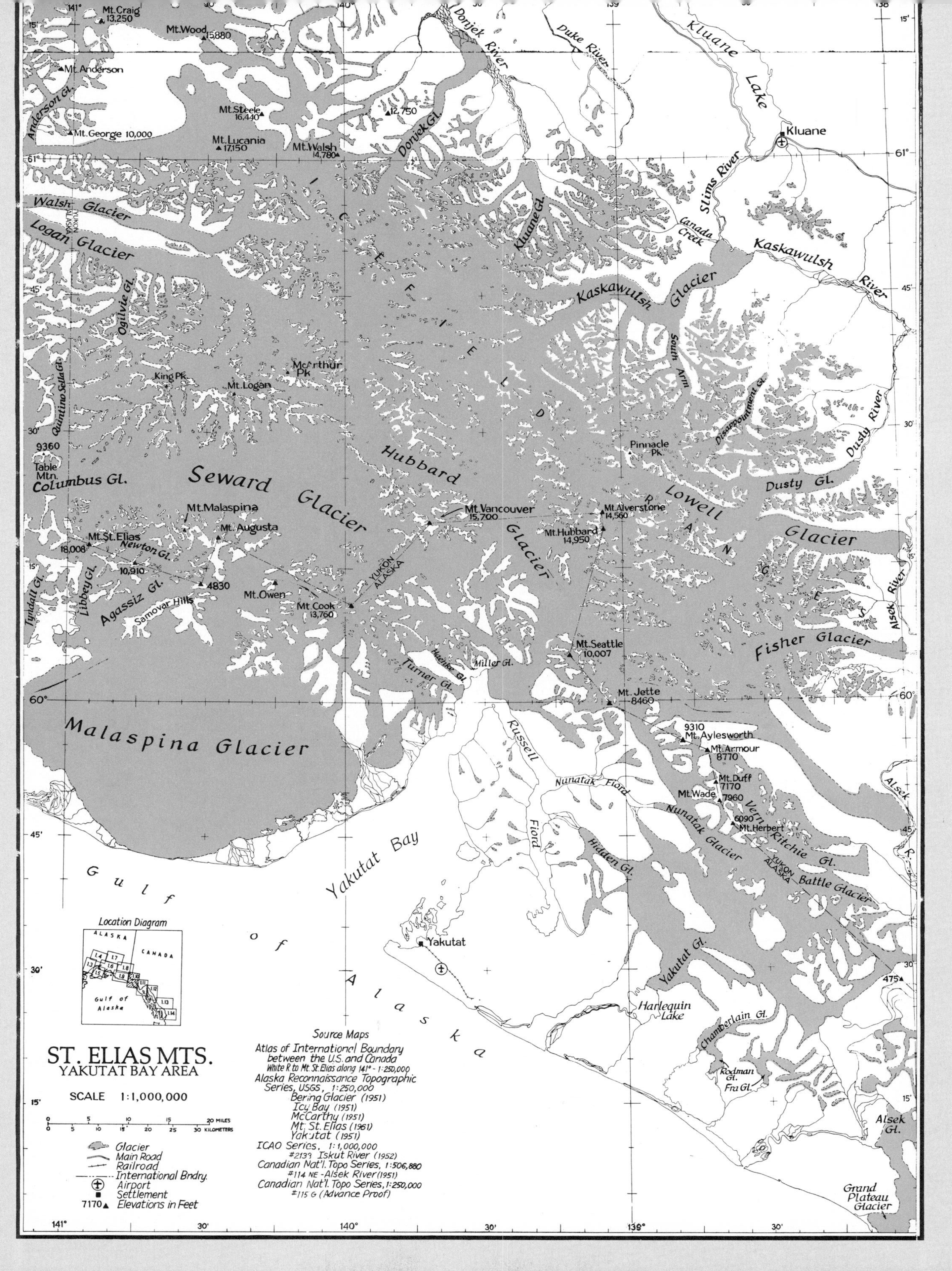

ST. ELIAS MTS.
YAKUTAT BAY AREA
SCALE 1:1,000,000
Location Diagram
ALASKA
CANADA
Gulf of Alaska
Glacier
Main Road
Railroad
International Bndry.
Airport
Settlement
7170 Elevations in Feet
Source Maps
Atlas of International Boundary between the U.S. and Canada
White R. to Mt. St. Elias along 141° - 1:250,000
Alaska Reconnaissance Topographic Series, USGS, 1:250,000
Bering Glacier (1951)
Icy Bay (1951)
McCarthy (1951)
Mt. St. Elias (1961)
Yakutat (1951)
ICAO Series, 1:1,000,000
#2133 Iskut River (1952)
Canadian Nat'l. Topo Series, 1:506,880
#114 NE - Alsek River (1951)
Canadian Nat'l. Topo Series, 1:250,000
#115 G (Advance Proof)
Kluane Lake
Kluane
Slims River
Kaskawulsh River
Kaskawulsh Glacier
South Arm
Canada Creek
Duke River
Donjek River
Donjek Gl.
Kluane Gl.
Dusty River
Dusty Gl.
Alsek River
Alsek R.
Alsek Gl.
Lowell Glacier
Fisher Glacier
Disappointment Gl.
Pinnacle Pk.
Mt. Alverstone 14,560
Mt. Hubbard 14,950
Mt. Seattle 10,007
Mt. Jette 8460
9310 Mt. Aylesworth
Mt. Armour 8770
Mt. Duff 7170
Mt. Wade 7960
6090 Mt. Herbert
Vern Ritchie Gl.
Battle Glacier
Nunatak Glacier
Nunatak Fiord
Hidden Gl.
Russell Fiord
Yakutat Gl.
Harlequin Lake
Chamberlain Gl.
Rodman Gl.
Fra Gl.
Grand Plateau Glacier
475
Yakutat
Yakutat Bay
Gulf of Alaska
Hubbard Glacier
Miller Gl.
Haenke Gl.
Turner Gl.
Mt. Vancouver 15,700
Seward Glacier
Malaspina Glacier
Mt. Cook 13,760
Mt. Owen
Mt. Augusta
Mt. Malaspina
Mt. St. Elias 18,008
Newton Gl.
10,910
4830
Agassiz Gl.
Samovar Hills
Libbey Gl.
Tyndall Gl.
Columbus Gl.
9360 Table Mtn.
Quintino Sella Gl.
Logan Glacier
Walsh Glacier
Ogilvie Gl.
King Pk.
Mt. Logan
McArthur Pk.
Mt. Walsh 14,780
Mt. Lucania 17,150
Mt. Steele 16,440
12,750
Mt. Wood 15,880
Mt. Craig 13,250
Mt. George 10,000
Mt. Anderson
Anderson Gl.
YUKON
ALASKA

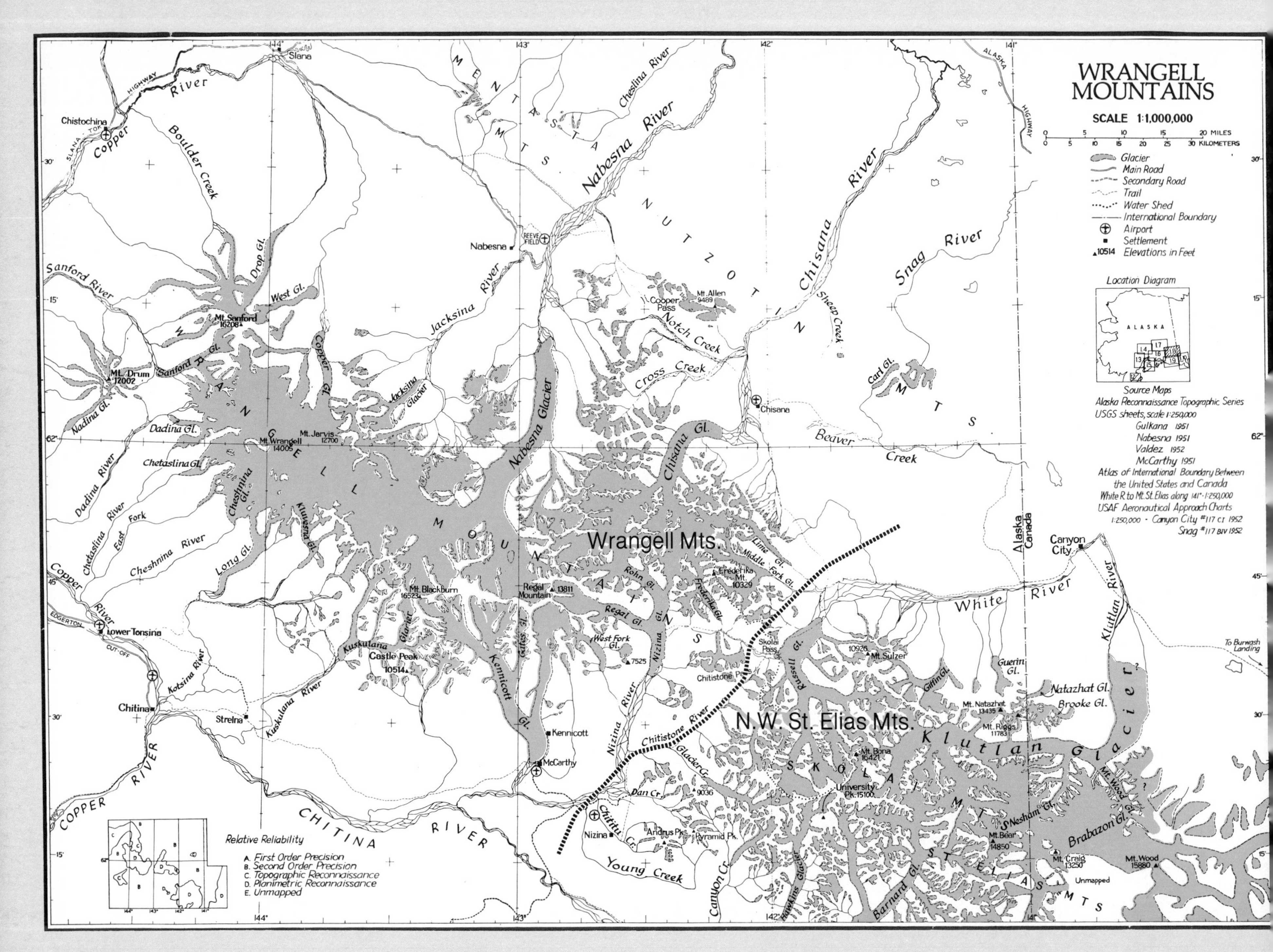
WRANGELL MOUNTAINS
SCALE 1:1,000,000
Glacier
Main Road
Secondary Road
Trail
Water Shed
International Boundary
Airport
Settlement
Elevations in Feet
Location Diagram
ALASKA
Source Maps
Alaska Reconnaissance Topographic Series
USGS sheets, scale 1:250,000
Gulkana 1951
Nabesna 1951
Valdez 1952
McCarthy 1951
Atlas of International Boundary Between the United States and Canada
White R. to Mt. St. Elias along 141° - 1:250,000
USAF Aeronautical Approach Charts
1:250,000 - Canyon City #117 CI 1952
Snag #117 BIV 1952
Relative Reliability
A. First Order Precision
B. Second Order Precision
C. Topographic Reconnaissance
D. Planimetric Reconnaissance
E. Unmapped
Wrangell Mts.
N.W. St. Elias Mts.
WRANGELL MOUNTAINS
MENTASTA MTS
NUTZOTIN MTS
SKOLAI MTS
ST. ELIAS MTS
Slana
Chistochina
Nabesna
Chisana
Canyon City
Lower Tonsina
Chitina
Strelna
Kennicott
McCarthy
Nizina
Mt. Sanford 16208
Mt. Drum 12002
Mt. Wrangell 14005
Mt. Jarvis 12700
Mt. Blackburn 16523
Regal Mountain 13811
Castle Peak 10514
Mt. Allen 9489
Mt. Frederika 10329
Mt. Sulzer
Mt. Natazhat 13435
Mt. Riggs 11783
Mt. Bona 16421
University Pk. 15100
Mt. Bear 14850
Mt. Craig 13250
Mt. Wood 15880
Nabesna Glacier
Chisana Gl.
Klutlan Glacier
Kennicott Gl.
Nabesna River
Chisana River
Copper River
Chitina River
White River
Snag River
Beaver Creek
Alaska Canada
Unmapped
To Burwash Landing

Below — *Bradford Washburn's 1938 photo of Barnard Glacier, in the Saint Elias Range, brought the glacier its fame. Located on the south side of Mount Bona, Barnard is well-known for its numerous straight, parallel medial moraines.* (Bradford Washburn)
Right — *A close-up look at Mount Saint Elias (18,008 feet) shows that the mountain's slopes are covered with small glaciers. This photo was taken from Libbey Glacier, at an elevation of approximately 3,000 feet.* (Gil Mull)

documented that the surge had started at least three years earlier. By the late 1960s, surge movement had ceased, but the glacier's spurt was remarkable both for the magnitude of ice movement and also for the duration of surging.

For the past 40 years photographs of Barnard Glacier have appeared in many books and travel articles, but a search of published literature could produce no information about studies of the glacier. Flowing from the south side of Mount Bona, Barnard Glacier (not to be confused with Barnard Glacier in College Fiord of Prince William Sound) is about 30 miles long with its terminus just north of the Chitina River. In 1938 Bradford Washburn made a spectacular photograph of the glacier that showed numerous straight, parallel

One of the most spectacular aspects of Malaspina Glacier, the largest piedmont glacier in North America, is its complex system of contorted, folded and twisted moraines. Surges within Malaspina and its tributaries have caused movement in the entire glacier system, giving the resulting moraines the appearance of an abstract painting.

medial moraines, many unwavering for distances of more than five miles. It is this photograph that has brought Barnard its fame.

GLACIERS OF YAKUTAT BAY: Yakutat Bay forms a 30-mile-long and 18-mile-wide indentation on the east side of Malaspina Glacier. Russell Fiord joins it at Disenchantment Bay, the narrow, northern neck of upper Yakutat Bay.

Perhaps as recently as 600 years ago, as a result of an advance of Hubbard Glacier that culminated between 970 and 1290 A.D., the entire length of Yakutat Bay was completely glacier covered. Curved ridges at Monti Bay and Yakutat are the terminal moraine and recessional moraines that mark this maximum ice advance.

Once ice retreat began, sometime before 1400, it proceeded slowly. An underwater moraine at the lower end of Disenchantment Bay resulted from a later glacial advance that culminated about 1700 and before 1791. The ice retreat may have continued until about 100 years ago. Since 1894, when systematic mapping of Hubbard's terminus began, the general trend has been one of slow re-advance.

Hubbard Glacier, the gigantic river of ice that presently stands at the head of Yakutat Bay, is the longest valley glacier in North America with a total length, including its Canadian source, of 92 miles. Hubbard's tidewater terminus is more than six miles long and reaches heights of close to 300 feet.

Much has been written about the possibility of Hubbard advancing and sealing the mouth of Russell Fiord. This is a distinct possibility as an advance of slightly more than one-half mile is all that is necessary. However, Hubbard does not show any evidence of immediately advancing over Osier Island, the first step in closing off Russell Fiord. Between 1895 and the early 1970s, the advance of Hubbard Glacier narrowed Russell Fiord's entrance

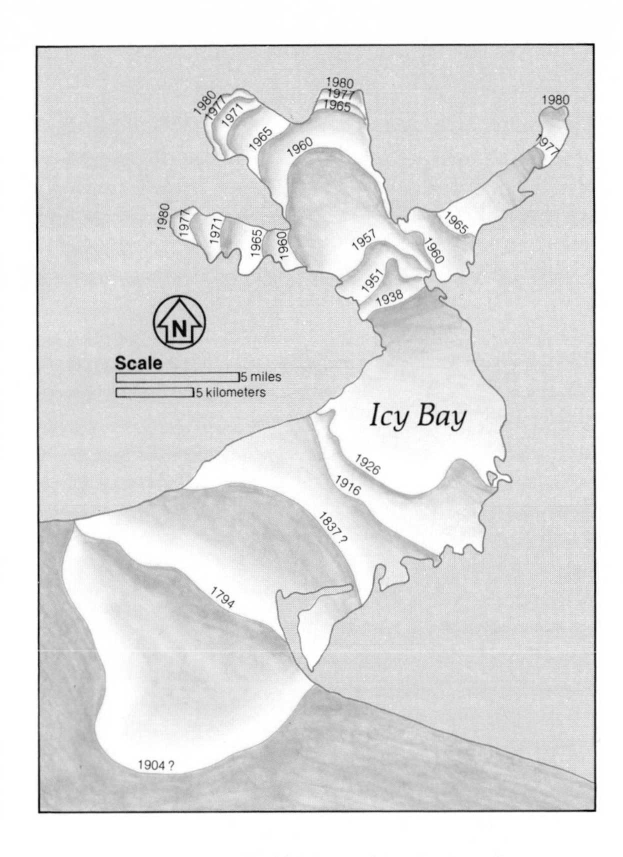

Figure 9 — *Deglaciation of Icy Bay on the northern Gulf of Alaska coast is rapidly continuing. Since 1904, more than 20 miles of the bay have been exposed.*

from more than two miles to less than one mile. During 1973 and 1974 a short-lived surge reduced this distance by another 600 feet. If the steady advance of Hubbard continues at the rate of about 100 feet per year, as it has for the past 75 years, then Russell Fiord may be sealed off in about 40 years. Geological evidence suggests that Russell Fiord may have been previously blocked sometime during the last 200 to 500 years.

Hidden Glacier in Russell Fiord has been retreating steadily from 1909 to the present with a total withdrawal of about four and one-quarter miles. However, Tarr and Martin documented that during the three years from 1906 to 1909, Hidden Glacier surged forward two miles. When first observed in 1891 Hidden Glacier was retreating, and continued to do so until the 1906 surge began.

Variegated Glacier, adjacent to but independent of Hubbard, also has a surge-dominated recent history. The glacier is about 15 miles long with a uniform slope of about five degrees from head to terminus. At least four times in the last 75 years Variegated Glacier has experienced significant surges. During the last surge, between the summers of 1964 and 1965, Variegated advanced nearly four miles. Previous surges were in 1905 and 1906, between 1911 and 1933, and just prior to 1948.

Intensive field investigation was undertaken by the Universities of Washington and Alaska, and the California Institute of Technology to try to determine what causes the surges which have been occurring approximately every 20 years. To date investigations have concentrated on thermal regime, mass balance, geometry, and motion of Variegated Glacier. A complete explanation for the surges has not emerged, but they seem related to the storage of water and its motion in the glacier.

William O. Field reports that Turner Glacier, on the west side of Disenchantment Bay, underwent catastrophic calving as the result of the July 9, 1958, earthquake centered near Lituya Bay. As much as 2,300 feet of Turner Glacier's tidewater terminus broke loose from the main body of the glacier in a series of seemingly endless calving events. Field suggests that the heavy calving may have resulted from submarine landslides beneath the ice mass which would reduce support and create an unstable condition.

Aside from the earthquake-induced changes, Turner Glacier has a long-term history of small fluctuations from its terminal moraine position. Austin Post reports that Turner Glacier has experienced a number of small surges.

Many other carefully studied glaciers exist in the Disenchantment Bay-Russell Fiord area. Haenke and Miller glaciers in Disenchantment Bay and Nunatak Glacier in Nunatak Fiord have been observed for the last 100 years — by Israel C. Russell in the early 1890s, the Harriman Alaska Expedition of 1899, and Tarr and Martin. The majority of these glaciers have been retreating.

MALASPINA GLACIER COMPLEX: The largest piedmont glacier in North America, Malaspina covers approximately 850 square miles between Yakutat Bay and Icy Bay at elevations of less than 2,000 feet. Malaspina Glacier measures about 45 miles from east to west and about 30 miles from north to south. The circumference of its **lobate terminus** is almost 60 miles. Frequent descriptions compare Malaspina to the size of Rhode Island.

More than 25 glaciers, the two largest being Seward, which extends into Canada, and Agassiz, supply ice to Malaspina. Including the total areas of all tributary glaciers, many of which are in Yukon Territory, Canada, the complete Malaspina Glacier complex has an area approaching 2,000 square miles.

Seismic studies conducted by Clarence Allen and George I. Smith in 1951 showed that the ice in the piedmont lobe attains a thickness of between 1,130 and 2,050 feet, and fills a channel or fiord to depths of 700 feet below sea level. Malaspina advanced to its maximum Little Ice Age position within the last two centuries; in late 1980, however, the glacier's terminus was melting and **downwasting.**

Stairway Icefall tumbles down a slope near the head of Root Glacier, located five miles north of the town of McCarthy. The glacier, named in 1899 for then Secretary of War Elihu Root, has very well defined lateral moraines and a surface crisscrossed with crevasses and wave ogives. (George Herben)

One of the most striking features of Malaspina Glacier is the complex of contorted, folded, and twisted moraines on the glacier's surface. Surges within the Malaspina system cause differential movement in the entire glacier system. Over the long period of time that the surging has occurred, the many moraines have taken on the appearance of an abstract painting.

About 30 years ago Malaspina Lake began to form in the stagnant ice of the glacier's southeast side. The lake is growing and now covers more than 20 square miles.

Many kettles, some covered by trees, are forming and expanding as the stagnant ice melts at Sitkagi Bluffs, which make up Malaspina's southern terminus. The largest individual kettles are more than 400 feet in diameter.

ICY BAY: Like Glacier Bay, the history of Icy Bay is also quite dynamic. As recently as 1904 Icy Bay did not exist. In 1794, when Vancouver explored the Gulf of Alaska coast, Guyot Glacier, then connected to Malaspina, extended to the shoreline and filled the basin of present-day Icy Bay. A second separate bay was located about five miles to the east. By 1837, when Sir Edward Belcher sailed along the Gulf of Alaska coast, the second bay had disappeared, filled in by sediment from Guyot Glacier, which had retreated, opening a small indentation at the mouth of the present bay. By 1886 Guyot Glacier had re-advanced to a position about six miles seaward of the present shoreline. A large terminal moraine, part of which lies on the floor of the Gulf of Alaska, marks this maximum advance. Local Indian legend tells of a Native village located on the west side of Icy Bay that was destroyed by an advance of a glacier. It was this 19th century advance, or perhaps an advance in the previous four centuries, that is responsible for the legend. Ice retreat began just after the start of the 20th century and still continues. By late 1980 the retreat had exceeded 25 miles, a true example of catastrophic retreat.

Today four separate fiords are located at the head of Icy Bay. Two are filled by arms of Guyot Glacier, one by Yahtse Glacier, and one by Tyndall Glacier. All of the fiords are lengthening as active ice retreat and calving continues. In the 1960s and 1970s, as ice retreat continued in the southwesternmost fiord, a number of small tributary fiords emerged. At present the lengthening of these small fiords has stopped as the ice has retreated to the edge of tidewater. Water depths in the fiords of upper Icy Bay exceed 500 feet.

THE SKOLAI MOUNTAINS: In the Skolai Mountains, between the Saint Elias and Wrangell mountains, are more than a dozen glaciers with lengths of five miles or greater. Largest of these, Russell, is 27 miles long. Skolai Pass, adjacent to Russell Glacier, was an important access route from the Chitina Valley to Alaska's interior during the turn of the century gold rush. Explorers also followed the Skolai route and hence many of the area's glaciers have long histories of observation.

A recent study by George Denton and Wibjorn Karlen of the University of Maine carefully examined the Holocene history of seven glaciers in the Skolai Mountains: Guerin, Giffin, Sheep, Russell, Moraine Creek, Natazhat, and an unnamed glacier between Giffin and Guerin. The area experienced a widespread glacial advance that culminated about 2,700 to 2,800 years ago, followed shortly thereafter by retreat. About 1,250 years ago a short-lived advance took place, followed by retreat about 1,050 years

Straight, parallel moraines are a feature of Kennicott Glacier, one of several large glaciers which descend from the snowfields of the eastern Wrangell Mountains. The nunatak at left center is known as Packsaddle Island. Snow-covered Mount Churchill (15,638 feet) and Mount Bona (16,421 feet), in the Saint Elias Range, are the two highest peaks on the skyline. (George Herben)

ago. During the Little Ice Age a series of advances occurred, the most recent being in the earliest part of this century. Denton and Karlen found twisted and bent pieces of an old horse carriage buried in the moraine of Russell Glacier at Skolai Pass. They used this and botanical evidence to conclude that this most recent advance occurred between 1881 and 1921. Retreat of all of the investigated glaciers has continued up to the present time.

Wrangell Mountains

The Wrangell Mountains, a large, young, volcanic massif with a length of about 100 miles and a width of 60 miles, are located between Nabesna to the north, the Saint Elias Range to the east, the Chitina River to the south, and the Copper River to the west. Mount Blackburn (16,390 feet), Mount Sanford (16,237 feet), Mount Wrangell (14,163 feet), and Mount Drum (12,010 feet) are the principal peaks in the area and are all either active or dormant volcanoes or are composed of layered volcanic material. Glaciers of the Wrangell Mountains cover more than 3,200 square miles and include a broad upland ice field that extends from Mount Blackburn to Mount Sanford, a distance of more than 25 miles. Among the valley glaciers that drain this area are Nabesna, longest inland glacier in North America, and Copper, Sanford, Cheshnina, Long, and Kuskulana glaciers. Other large glaciers that descend from snowfields in the eastern Wrangells are Chisana, Nizina, and Kennicott glaciers.

Mount Wrangell is the only currently active volcano in the Wrangells. Studies by Dr. Carl Benson and others from the University of Alaska have shown that Mount Wrangell's heat flow has increased markedly during the past 15 years. Glacial and volcanic research, which began in 1960, has attempted to measure the effect of increasing volcanic heat flow on the glacier ice. Spectacular changes have occurred at the volcano's summit. About 65 million cubic yards of ice has melted from the interior and immediate surrounding area of one of the three active craters along the rim of Mount Wrangell's summit caldera. Mount Wrangell's glaciers are being used as a huge natural "calorimeter" — the melting rate of glacier ice is being measured annually as a means of estimating the volcanic heat flow.

Many other glaciers in the Wrangells have been investigated in the past. Nabesna, about 50 miles long and covering nearly 400 square miles, may have as many as 40 tributaries. In spite of its enormity, aerial photographs suggest that Nabesna Glacier is retreating.

Chisana, second largest glacier in the Wrangells with a length of almost 30 miles and an area of about 140 square miles, was retreating during the first half of this century. Since the 1950s, the glacier's terminus has fluctuated and the glacier has experienced a number of small surges. Austin Post, who has photographed the Wrangell Mountains glaciers annually for the past 20 years, reports that as of late 1980 three or four recent moraines are located adjacent to Chisana Glacier's active ice margin.

Chugach and Kenai Mountains

The Chugach and Kenai mountains, which arc around the northern Gulf of Alaska, comprise about one-third of Alaska's glaciated area, or more than 10,000 square miles. The mountains extend for more than 500 miles and contain: Bering Glacier, with the second largest piedmont lobe in Alaska; the fiord glaciers of Prince William Sound, including Columbia Glacier; and the Sargent and Harding icefields.

Eastern Chugach Mountains

Bering Glacier, largest in continental North America, consists of a piedmont lobe; a valley glacier section; and immense Bagley Icefield, which occupies a linear trench from north of the Icy Bay-Mount Saint Elias region to southwest of the Tana River, thus bridging the Saint Elias and eastern Chugach Mountains.

Bering Glacier's piedmont lobe is 25 miles across from east to west and 25 miles north to south, from the Khitrov Hills to Seal River. The distance around the perimeter, from Bering River to Grindle Hills, is about 47 miles.

On the west side of Bering Glacier kettles and vegetation are developing on the contorted moraines.

Including all its tributary glaciers, Bering's total area is about 2,250 square miles, about 250 square miles larger than the Seward-Malaspina system. Bering's piedmont lobe is about one-third smaller than that of Malaspina Glacier, however. Melting along Bering's terminus has resulted in development of a series of lakes along much of the big glacier's perimeter. Large icebergs are calving directly into these lakes from the stagnating terminus.

Between 1959 and 1960, and again between 1965 and 1966, Bering Glacier experienced a series of surges that further folded its medial moraines and displaced ice as much as eight miles. In addition to the internal deformation, the terminus of Bering Glacier advanced at a few locations.

Lower Copper River Region

The Copper River, which originates at Copper Glacier on the northeast flank of Mount Wrangell, is the largest river draining into the Gulf of Alaska, and may carry more sediment, almost all of which is glacially derived, than any other river in Alaska, including the Yukon. About a dozen large glaciers either comprise part of the shoreline of the lower Copper River, drain into it, or drain into its delta.

Many of these glaciers have played an important role in complicating the economic development of the area. Around the turn of the century, when large quantities of copper were discovered at Kennecott in the Wrangells, a route had to be found to transport the copper ore to the port of Cordova, from where the ore could be shipped to smelters in Tacoma, Washington, and elsewhere. The selected route followed the Copper River Valley, and had to pass by Heney, Allen, Childs, Grinnell, and Miles glaciers. Between 1906 and 1910 the Copper River & Northwestern Railway was constructed at a cost of about $20 million. What made the railroad's construction remarkable is that track was laid on the surface of two of the glaciers: on five and one-half miles of Allen Glacier, and one-quarter mile of moraine-covered ice of Grinnell Glacier; and on about one-half mile of terminal moraine of Heney Glacier. Construction crews lived in fear of an advance of the other glaciers and of changes in the glaciers on which track was laid. Workers continuously struggled to maintain the track and compensate for glacier movements.

Miles Glacier empties directly into the Copper River at Miles Lake, an almost circular body of water about four miles in diameter. Large numbers of icebergs calve off this glacier's terminus and float across the lake, passing under the ruins of the Million Dollar Bridge (partially destroyed by the 1964 earthquake) as they drift downstream. To protect the bridge from Miles Glacier's icebergs, massive steel and concrete iceberg deflectors were installed in front of the bridge's main supports when it was built in 1909.

In the last 150 years Miles has retreated almost four miles. This retreat, however, has been interrupted by two small periods of advance, 1885 to 1888 and 1908 to 1910. In 1888 the terminus was only 400 feet from the site of the Million Dollar Bridge. Much stagnant ice from the terminus of the early 20th century advance is located on the west side of the Copper River, having been separated by the river in the past 50 years.

Tarr and Martin reported that in July 1909 a section of Miles Glacier about one-half mile square broke away from the terminus and produced icebergs that completely filled Miles Lake. Today Miles is still an active iceberg producer.

Childs Glacier, immediately west of the Million Dollar Bridge, is 12 miles long and covers more than 40 square miles. The glacier forms the west shoreline of the Copper River for about four miles. Between the spring of 1909 and summer of 1911, during construction of the Million Dollar Bridge, Childs Glacier advanced from a position about 3,440 feet from the bridge to a position only 1,474 feet from the bridge. The majority of this advance, 1,665 feet, occurred between March 1909 and June 3, 1910. The average rate of advance during this period was about four feet per day. During late summer of 1910 a short-lived surge exceeded eight feet per day. Fortunately for the bridge, Childs Glacier's advance stopped in 1912 and the ice has not threatened the bridge since.

Allen Glacier, seven miles upriver from Childs, is about 20 miles long and terminates at the Copper River. Unlike Childs and Miles glaciers, which have iceberg-calving termini along the Copper River shoreline, Allen's terminus is moraine-covered stagnant ice. During construction of the railroad, the railbed was built for five and one-half miles on what Tarr and Martin described in *Alaskan Glacier Studies* (1914) as "the end of a living glacier," and they observed "that the ballast beneath the ties and rails of the railway actually rests upon the ice, not upon an abandoned moraine as at Heney Glacier."

Tarr and Martin pointed out that problems encountered included "the settling of the track as the ice beneath it melts; the frequent breaking out of new streams" thus requiring new trestles; and the 18-inch shifting of a bridge support. Advances of both Allen

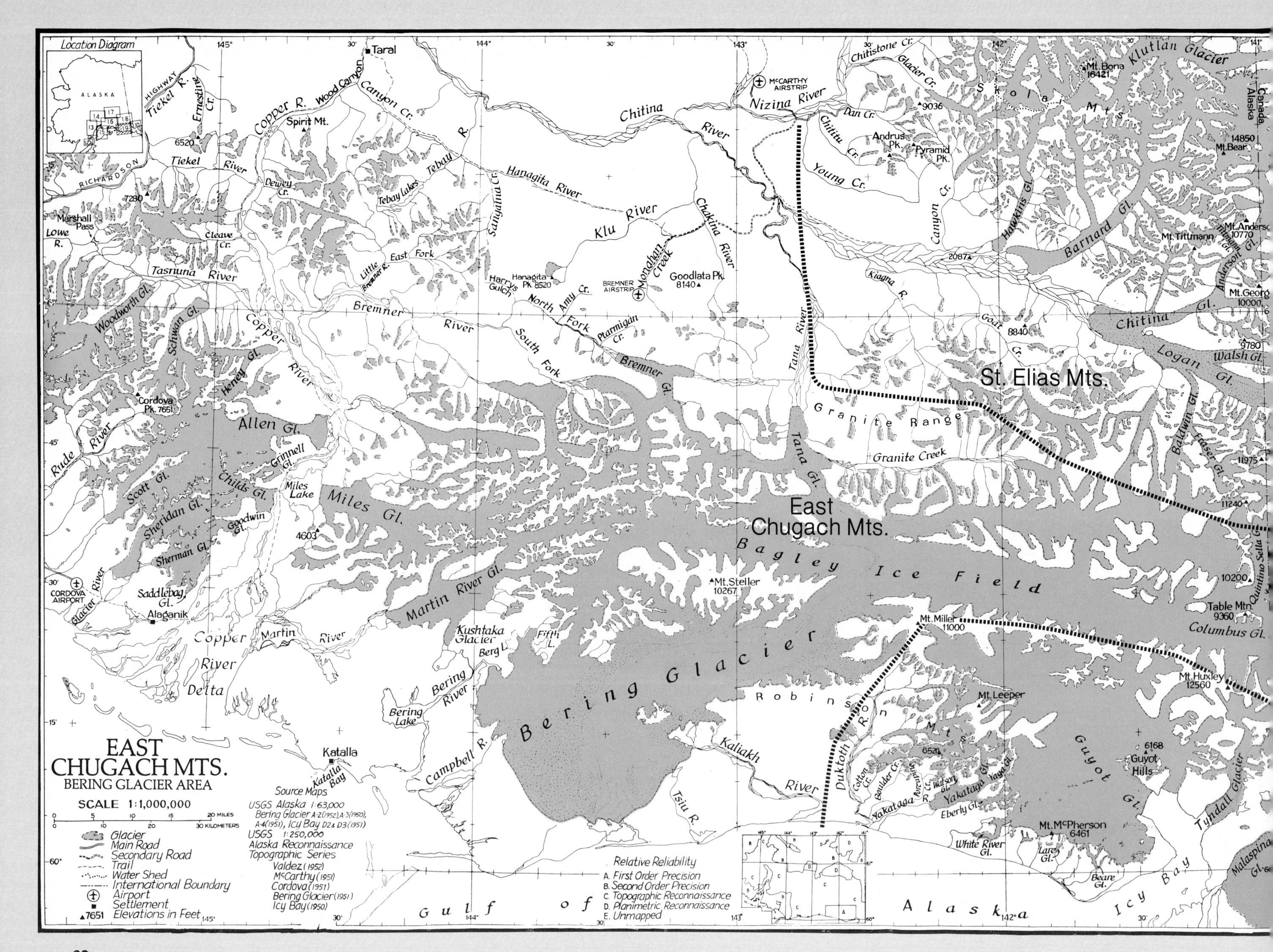
EAST
CHUGACH MTS.
BERING GLACIER AREA
SCALE 1:1,000,000
Location Diagram
ALASKA
Glacier
Main Road
Secondary Road
Trail
Water Shed
International Boundary
Airport
Settlement
Elevations in Feet
Source Maps
USGS Alaska 1:63,000
USGS 1:250,000
Alaska Reconnaissance
Topographic Series
Valdez (1952)
McCarthy (1951)
Cordova (1951)
Bering Glacier (1951)
Icy Bay (1950)
Relative Reliability
A. First Order Precision
B. Second Order Precision
C. Topographic Reconnaissance
D. Planimetric Reconnaissance
E. Unmapped
East
Chugach Mts.
St. Elias Mts.
Bagley Ice Field
Bering Glacier
Copper R.
Taral
Chitina River
Nizina River
McCARTHY AIRSTRIP
Hanagita River
Klu River
Bremner River
Tasnuna River
Tiekel River
Allen Gl.
Miles Gl.
Miles Lake
Martin River Gl.
Copper River Delta
Cordova Pk. 7651
CORDOVA AIRPORT
Alaganik
Katalla
Bering Lake
Bering River
Kaliakh River
Duktoth R.
Robinson Mts.
Mt. Steller 10267
Mt. Miller 11000
Mt. Huxley 12560
Mt. Leeper
Mt. McPherson 6461
Guyot Hills
Guyot Gl.
Columbus Gl.
Tyndall Glacier
Granite Range
Granite Creek
Tana Gl.
Tana River
Chitina Gl.
Logan Gl.
Barnard Gl.
Mt. Bona 16421
Klutlan Glacier
Skolai Mts.
Mt. Bear
Mt. Tittmann
Goodlata Pk. 8140
BREMNER AIRSTRIP
Hanagita Pk. 8520
Kushtaka Glacier
Gulf of Alaska
Icy Bay
Malaspina Gl.
Canada
Alaska

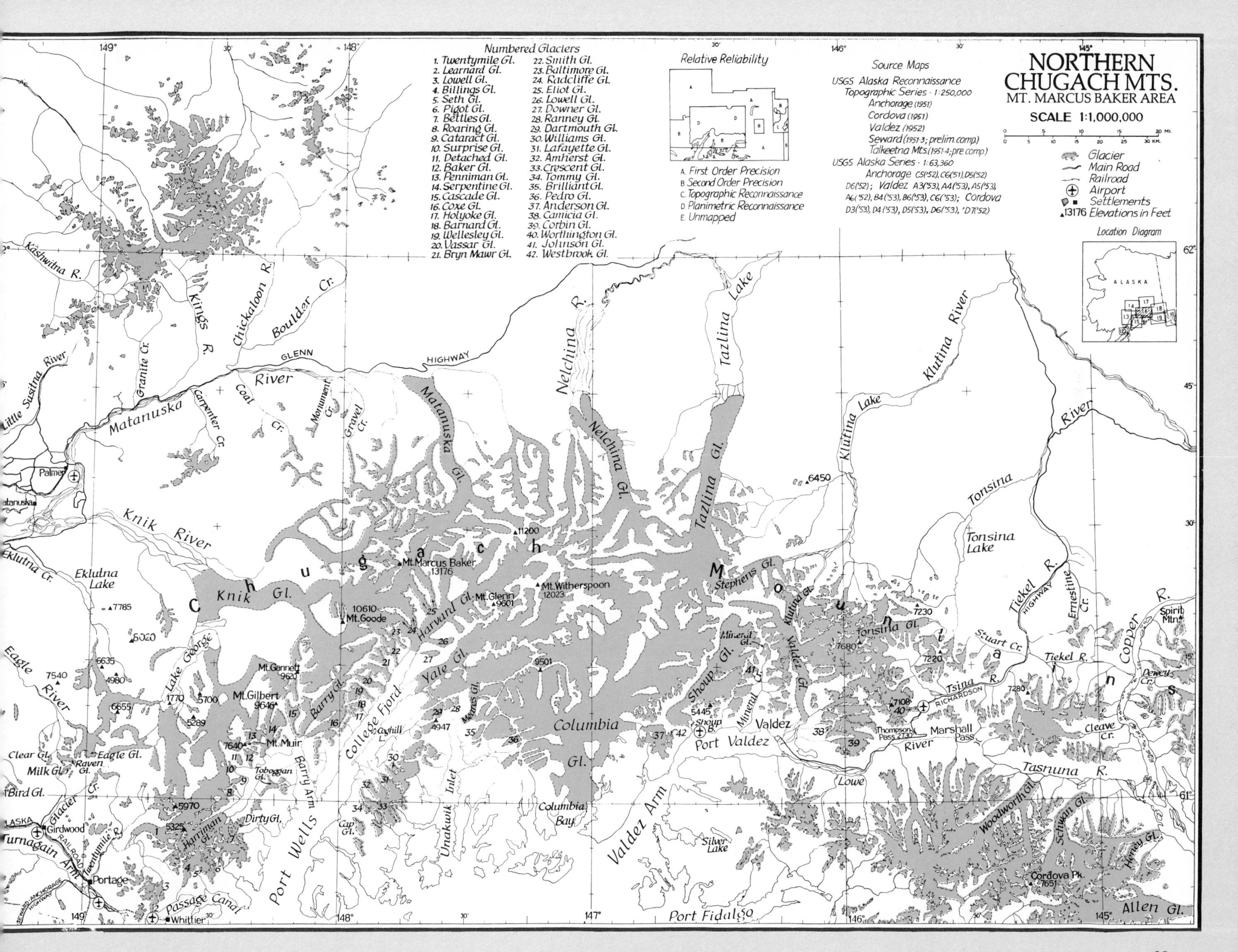
NORTHERN CHUGACH MTS.
MT. MARCUS BAKER AREA
SCALE 1:1,000,000
Numbered Glaciers
1. Twentymile Gl.
2. Learnard Gl.
3. Lowell Gl.
4. Billings Gl.
5. Seth Gl.
6. Pigot Gl.
7. Bettles Gl.
8. Roaring Gl.
9. Cataract Gl.
10. Surprise Gl.
11. Detached Gl.
12. Baker Gl.
13. Penniman Gl.
14. Serpentine Gl.
15. Cascade Gl.
16. Coxe Gl.
17. Holyoke Gl.
18. Barnard Gl.
19. Wellesley Gl.
20. Vassar Gl.
21. Bryn Mawr Gl.
22. Smith Gl.
23. Baltimore Gl.
24. Radcliffe Gl.
25. Eliot Gl.
26. Lowell Gl.
27. Downer Gl.
28. Ranney Gl.
29. Dartmouth Gl.
30. Williams Gl.
31. Lafayette Gl.
32. Amherst Gl.
33. Crescent Gl.
34. Tommy Gl.
35. Brilliant Gl.
36. Pedro Gl.
37. Anderson Gl.
38. Camicia Gl.
39. Corbin Gl.
40. Worthington Gl.
41. Johnson Gl.
42. Westbrook Gl.
Relative Reliability
A. First Order Precision
B. Second Order Precision
C. Topographic Reconnaissance
D. Planimetric Reconnaissance
E. Unmapped
Source Maps
USGS Alaska Reconnaissance
Topographic Series - 1:250,000
Anchorage (1951)
Cordova (1951)
Valdez (1952)
Seward (1951-3; prelim. comp.)
Talkeetna Mts. (1951-4; pre comp.)
USGS Alaska Series - 1:63,360
Anchorage C5('52), C6('51), D5('52)
D6('52); Valdez A3('53), A4('53), A5('53),
A6('52), B4('53), B6('53), C6('53); Cordova
D3('53), D4('53), D5('53), D6('53), D7('52)
Glacier
Main Road
Railroad
Airport
Settlements
13176 Elevations in Feet
Location Diagram
ALASKA
Chugach Mountains
Mt. Marcus Baker 13176
Mt. Witherspoon 12023
Mt. Glenn 9601
Mt. Goode 10610
Mt. Gannett 9620
Mt. Gilbert 9646
Mt. Muir
Knik Gl.
Harvard Gl.
Yale Gl.
Barry Gl.
Columbia Gl.
Matanuska Gl.
Nelchina Gl.
Tazlina Gl.
Stephens Gl.
Klutina Gl.
Tonsina Gl.
Valdez Gl.
Shoup Gl.
Mineral Gl.
Woodworth Gl.
Schwan Gl.
Heney Gl.
Allen Gl.
Harriman Gl.
Dirty Gl.
Toboggan Gl.
Cap Gl.
Eagle Gl.
Raven Gl.
Clear Gl.
Milk Gl.
Bird Gl.
Glacier Cr.
GLENN HIGHWAY
RICHARDSON HIGHWAY
Kashwitna R.
Kings R.
Chickaloon R.
Boulder Cr.
Little Susitna River
Granite Cr.
Matanuska River
Carpenter Cr.
Coal Cr.
Monument Gl.
Gravel Cr.
Nelchina R.
Tazlina Lake
Klutina Lake
Klutina River
Tonsina
Tonsina Lake
Tiekel R.
Ernestine Cr.
Copper R.
Spirit Mtn.
Stuart Cr.
Dewey Cr.
Cleave Cr.
Tsina R.
Tasnuna R.
Lowe River
Marshall Pass
Thompson Pass 2730
Palmer
Matanuska
Knik River
Eklutna Cr.
Eklutna Lake
Eagle River
Lake George
College Fjord
Barry Arm
Coghill L.
Unakwik Inlet
Columbia Bay
Port Wells
Valdez Arm
Port Valdez
Valdez
Shoup B.
Silver Lake
Port Fidalgo
Cordova Pk. 7651
Passage Canal
Whittier
Portage
Girdwood
Turnagain Arm
Twentymile R.
ALASKA RAILROAD
SEWARD-ANCHORAGE HIGHWAY
Mt. 11200
6450
7785
6020
6635
4980
7540
6655
1770
5700
5289
7640
5970
5325
9501
4947
5445
7680
7230
7220
7108
7280
149°
148°
147°
146°
145°
62°
61°

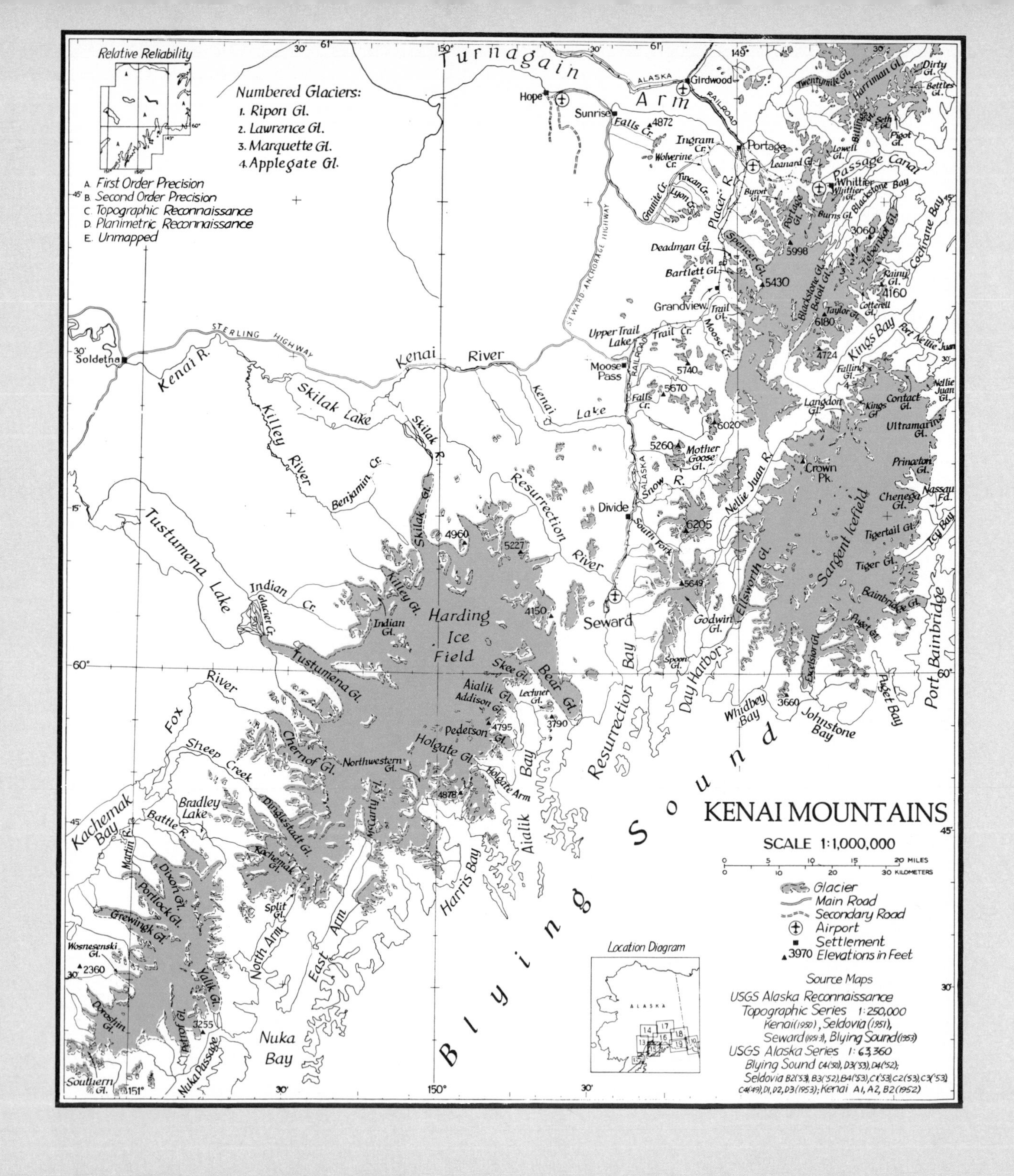

KENAI MOUNTAINS
SCALE 1:1,000,000
0 5 10 15 20 MILES
0 10 20 30 KILOMETERS
Glacier
Main Road
Secondary Road
Airport
Settlement
3970 Elevations in Feet
Source Maps
USGS Alaska Reconnaissance Topographic Series 1:250,000
Kenai (1950), Seldovia (1951), Seward (1951), Blying Sound (1953)
USGS Alaska Series 1:63,360
Blying Sound C4('50), D3('53), D4('52);
Seldovia B2('53), B3('52), B4('53), C1('53), C2('53), C3('53), C4('49), D1, D2, D3 (1953); Kenai A1, A2, B2 (1952)
Location Diagram
ALASKA
Relative Reliability
A. First Order Precision
B. Second Order Precision
C. Topographic Reconnaissance
D. Planimetric Reconnaissance
E. Unmapped
Numbered Glaciers:
1. Ripon Gl.
2. Lawrence Gl.
3. Marquette Gl.
4. Applegate Gl.
Turnagain Arm
Hope
Sunrise
Girdwood
Portage
Whittier
Passage Canal
Seward
Moose Pass
Divide
Soldotna
Grandview
Upper Trail Lake
Kenai R.
Kenai River
Kenai Lake
Skilak Lake
Killey River
Tustumena Lake
Harding Ice Field
Sargent Icefield
Resurrection River
Resurrection Bay
Blying Sound
Kachemak Bay
Nuka Bay
Aialik Bay
Harris Bay
Day Harbor
Port Bainbridge
Kings Bay
Cochrane Bay
Whidbey Bay
Johnstone Bay
Puget Bay
Crown Pk.
Tustumena Gl.
Skilak Gl.
Bear Gl.
Holgate Gl.
Chernof Gl.
Dinglestadt Gl.
Grewingk Gl.
Portlock Gl.
Dixon Gl.
Yalik Gl.
Spencer Gl.
Ellsworth Gl.
Bainbridge Gl.
Tiger Gl.
Chenega Gl.
Princeton Gl.
Ultramarine Gl.
Contact Gl.
Nellie Juan Gl.
Harriman Gl.
Twentymile Gl.
STERLING HIGHWAY
SEWARD ANCHORAGE HIGHWAY
ALASKA RAILROAD
4872
5998
5430
4160
6180
4724
5740
5670
6020
5260
6205
5649
4960
5227
4150
4795
3790
4878
3660
3060
2360
3255

Glacier, in 1912, and Grinnell Glacier, between 1909 and 1911, did occur but had no effect on the railroad, thus making its maintenance across the glaciers relatively easy. It is possible that construction of the Copper River & Northwestern Railway is the only instance where man has successfully constructed a railroad over glacier ice. What is most remarkable is that the railroad was maintained in operating condition for more than 20 years.

Sherman and Sheridan glaciers both head in the Chugach Mountains, west of the Copper River, and both drain into the Gulf of Alaska by way of the Glacier River.

The March 27, 1964, Alaska earthquake generated many large rockslides and avalanches that fell on surfaces of Chugach Mountain glaciers. Among the largest of these was a gigantic landslide that fell on Sherman Glacier, covering an area of about three square miles with more than 25 million cubic yards of rock debris. The debris, with maximum thickness of more than 25 feet, came from one mountain, Shattered Peak. Ronald Shreve of the University of California at Los Angeles examined the mechanics of this slide and concluded that the debris had moved downglacier on a cushion of air with velocities of up to 50 miles per hour. Part of the debris glided over a ridge about 500 feet high. Colin Bull and Cedomir Marangunic of Ohio State University noted that in the year following the slide, the ice surface below the slide area was lowered between 26 and 33 feet. They attributed much of this loss to abnormally high air temperatures that resulted from the heating of downglacier air as it passed over the debris' surface. While heat reflected by the slide may have contributed to increased melting in the bare ice downglacier, Bull postulated in 1968 that the debris cover would lower the rate of melting of the ice beneath the debris and result in a surge of Sherman Glacier in 1978 to 1980. By late 1980 there was no evidence of the predicted surge.

In 1965 two very large rockslides, longer than any that took place during the 1964 earthquake, occurred on Allen and Fairweather glaciers. The Allen Glacier slide consists of material that was probably loosened by the 1964 earthquake. However, the Fairweather slide occurred 360 miles to the southeast of the quake's epicenter and probably is not related to it. The 1965 slide on Fairweather, which straddles Fairweather Fault,

Six glaciers are visible in this photo of College Fiord. The glaciers are, from left, Bryn Mawr, Smith, Radcliffe, Harvard, Eliot (non-tidewater), and Yale. (Jon R. Nickles)

may have resulted from local movements of the fault. At least 25 other glaciers experienced 1964 earthquake-generated rockslides or avalanches. But the 1964 quake has not been the only cause of rockslides on Alaska's glaciers. Austin Post reported 11 glaciers that experienced rockslides or avalanches between 1945 and 1963, the largest of which occurred on Barry Glacier in 1960.

Prince William Sound

Prince William Sound is an intricate complex of fiords and islands that contains the greatest concentration of tidewater-calving glaciers anywhere in Alaska. More than 40 presently glaciated or recently glaciated fiords are cut into the margins of the Prince William Sound Basin; about one-third of the fiords contain the sound's 20 active tidewater glaciers. The fiords with active glaciers are located on the west and northwest sides of the sound and include: College Fiord, Harriman Fiord, Columbia Bay, Blackstone Bay, Barry Arm, Unakwik Inlet, Port Bainbridge, Port Nellie Juan, Icy Bay, and Nassau Fiord. In addition to the glaciers in Prince William Sound's fiords, many glaciers exist on Montague Island and at least one small glacier is located on Knight Island. The existence of glaciers on Chenega and Hinchinbrook islands was discussed by Tarr and Martin in 1914. Whether those glaciers still exist today is not known, but none are shown on the most recent topographic maps of the area.

College Fiord, located at the north end of Port Wells, is about 25 miles long and three miles wide and contains six calving tidewater glaciers: Harvard and Yale glaciers, in two arms at the fiord's head; and Smith, Bryn Mawr, Vassar, and Wellesley glaciers on its west wall. Named non-tidewater glaciers in College Fiord are Barnard, Holyoke, Radcliffe, Amherst, Dartmouth, Lafayette, Baltimore, Eliot, and Crescent. Except for Crescent, which was named for its shape, these glaciers were named by the geologists of the Harriman Alaska Expedition of 1899 for colleges and universities with which they had affiliations. Included are the names of all of the "Seven Sisters" women's colleges associated with the then all male Ivy League colleges, about one-half of the Ivy League Colleges, two finishing schools and two past presidents of Harvard University. In 1908, when Grant and Higgins gave the name Muth Glacier to a previously unnamed glacier, the uniqueness of the names of College Fiord glaciers was ended.

Most of College Fiord's glaciers have shown a general trend of slow minor recession since their mapping at the turn of the century. Harvard Glacier, however, has been slowly advancing for the past 75 years; and in recent years Yale Glacier has retreated rapidly, receding more than three miles during this century. The 1964 earthquake caused rockslides and avalanches on Smith, Vassar, Harvard, and Yale glaciers.

A recently published University of Alaska study by D. Van Wormer and E. Berg reported that as many as 60 distinct signals per day were being detected on seismographs around southcentral Alaska. Plots of epicenters of the signals, many with Richter magnitudes between 2.0 and 2.5, showed them to be generally located in College Fiord, on or near Harvard Glacier, the fiord's most active glacier. The report concludes that the seismic events are associated with the glaciers and that the energy of the larger events and the great number of frequent smaller events suggest that jerky, short-lived rapid motions are an important part of total glacier movement.

In 1899, when the Harriman Alaska Expedition arrived, upper Barry Arm, located at the northwest end of Port Wells, was almost completely filled by Barry Glacier. By sailing close to the terminus the expedition was able to pass around Barry Glacier and discover a completely unknown inlet, which they named Harriman Fiord. The 1899 ice position of Barry Glacier represents its maximum position for the last five centuries or more. Between 1899 and 1914, the glacier retreated more than four miles. Its position has changed little since then. Today, three glaciers are located at the head of Barry Arm: Barry, Cascade and Coxe glaciers. Coxe Glacier separated from retreating Barry Glacier between 1910 and 1914. Barry and Cascade glaciers almost separated during the retreat, but a small advance has since rejoined their termini.

Harriman Fiord contains more than a dozen glaciers, the longest being Harriman Glacier at its head. Most of the fiord's glaciers have retreated during this century, except for Harriman Glacier, which has been slowly advancing for at least the past 40 years. Surprise Glacier, so named because it was the first glacier seen when the Harriman Expedition entered this fiord, retreated a little more than a mile between 1899 and 1910. In late 1980 its position appeared to be stable.

Unakwik Inlet is a 20-mile-long, northward-trending fiord between College Fiord and Columbia Bay. Meares Glacier, studied by Grant and Higgins in 1905 and 1909, Lawrence Martin in 1910, and more recently by a host of others, is an iceberg-calving, tidewater glacier located at the fiord's north end. Meares Glacier has a history of advance from its first observations in the early 1900s through the present. At least twice during this

Left — *The behavior of Columbia Glacier has been extensively studied because of the glacier's proximity to the port of Valdez, terminus of the trans-Alaska oil pipeline. Increased iceberg production, as has been predicted by the U.S. Geological Survey, could pose a hazard to tankers carrying petroleum from Valdez.* (Jon R. Nickles)

Below — *The steamship* Yukon *carries passengers and freight past the terminus of Columbia Glacier in this photo, probably taken during the late 1920s. The* Yukon, *which began its service in Alaska in 1924, was destroyed in a tragic accident in 1946 when it ran aground and broke up in stormy seas near Seward.* (Tom Sexton)

cycle of advance there were periods when the glacier's terminal position remained stable. By late 1980 total advance had been about one-half mile. At present Meares Glacier's terminus is slowly advancing into the fiord in an area where water depths are greater than 500 feet. A large terminal moraine about halfway between the glacier and the fiord's mouth represents the maximum Neoglacial advance of Meares Glacier and makes navigation of Unakwik Inlet impossible for all but the smallest of boats.

Columbia Glacier, largest in Prince William Sound, was named by the Harriman Alaska Expedition for Columbia University. The glacier is more than 40 miles long, covers an area of about 440 square miles, and terminates at the head of Columbia Bay, a fiord with depths greater than 600 feet. Columbia's tidewater terminus, more than six miles long, has produced numerous icebergs in past years. The glacier had part of its terminus grounded on Heather Island in Columbia Bay between 1910 and early 1979, when the glacier retreated off the island.

Because of the proximity of Columbia

Knik Glacier, located in the Matanuska Valley north of Anchorage, advanced every winter between 1915 and 1966, blocking meltwater flow from several surrounding glaciers. Each spring the meltwater would form a lake behind the terminus of Knik Glacier, which, by early summer would overflow, causing a jokulhlaup. *The glacier is currently being studied by the U.S. Geological Survey in an attempt to determine why this annual occurrence ceased in 1966.* (John Koranda)

Glacier to the port of Valdez, terminus of the trans-Alaska oil pipeline, much research was conducted recently to determine what changes might occur in the glacier's terminus in the next few decades. Results of the research are not encouraging. In late 1979 the U.S. Geological Survey predicted that an "irreversible, drastic retreat of Columbia is inevitable" and that it "probably will begin in less than 20 years, and may already be in its early phase." The prediction further states that iceberg production will increase to four times the number of bergs produced in 1977-1978, and that ultimately a new, 25-mile-long fiord will open.

Further predictions regarding Columbia Glacier suggest that 20 to 27 million tons of ice will be produced from 1982 to 1985; retreat of up to five miles will occur by 1986, releasing up to 10 cubic miles of ice; and as many as 50 cubic miles of icebergs could be released into Prince William Sound during the next 30 to 50 years.

If retreat begins and iceberg production significantly increases, tankers carrying petroleum from Valdez to refineries in the lower 48 states may be restricted in terms of when and where they can travel in Prince William Sound. It is possible that Port Valdez could be completely blocked by icebergs, or in the worst possible case, a tanker-iceberg collision leading to an oil spill could occur. At times, the U.S. Coast Guard has monitored the drifting of Columbia Glacier icebergs out of Columbia Bay and has regulated ship traffic to avoid such accidents. However, no ongoing program presently exists to study the iceberg problem.

Valdez Glacier, about four miles above the pre-1964 townsite of Valdez, has a recent history closely tied to the turn of the century gold rush. The glacier served as a major highway from the port of Valdez to the gold fields of the interior. Valdez Glacier was used to bypass the rugged Chugach Mountains route through Keystone Canyon and Thompson Pass. A relatively easy, 18-mile ice route led to the head of Klutina Glacier and from there to Klutina Lake.

After 1910, mining took place on both walls of Valdez Glacier valley. William O. Field reports that access to these claims was by pack train and foot up the lower four to five miles of Valdez Glacier. In recent years thinning of the glacier has left these claims more than 300 feet above the present ice surface. Retreat continues at Valdez Glacier, which receded about three-fourths of a mile between 1901 and 1971.

Blackstone Glacier and Bay were named in 1899 for a miner who lost his life on the glacier in 1896. Like College Fiord, many of the glaciers in Blackstone Bay, south of Whittier, are also named for universities and colleges. In this instance, they were named in 1910 by Lawrence Martin for schools in Wisconsin. Glaciers in Blackstone Bay include Ripon, Lawrence, Marquette, Beloit, and Northland. A conspicuous underwater terminal moraine, part of which is exposed under most tidal conditions, connects Willard Island in the middle of the bay with the east shore of Blackstone Bay. Grant and Higgins suggest that the moraine predates the early 18th century. Today, Blackstone and Beloit glaciers are slowly advancing, while all of the other glaciers in Blackstone Bay appear to be retreating.

The source of ice for most of the glaciers southwest of Blackstone Bay is an unnamed ice field covering an area of almost 200 square miles. Aside from the glaciers of

Blackstone Bay, other major glaciers that drain from this ice field include Whittier, Portage, Spencer, Bartlett, Taylor, Tebenkof, and Wolverine glaciers, all of which have retreated appreciable distances since first observed. Wolverine Glacier has been monitored continuously by the U.S. Geological Survey since 1966.

Portage Glacier is the most popular glacier in the Anchorage area and the second most visited glacier in Alaska. For centuries Portage served as an overland passage, a portage, between Turnagain Arm and Prince William Sound. Portage Glacier stood at its recent maximum position in 1880, when it began to retreat, and by 1914 had exposed the northwest end of Portage Lake Basin. To date, retreat is slightly more than two miles. Continued retreat at the present rate would lengthen the lake by an additional mile and put the terminus at a point of stability by the year 2020 according to a prediction by Larry R. Mayo, Chester Zenone, and Dennis Trabant of the U.S. Geological Survey. This would leave the terminus out of sight of the point from which it can presently be viewed, the U.S. Forest Service Visitor Center. Today Portage Glacier is known for the tremendous accumulation of icebergs that drift to the shoreline in front of the visitor center. With continued retreat, the glacier may disappear from view, but it is thought that the spectacular iceberg production would continue.

Northern and Western Chugach Mountains

Many glaciers in northwestern Prince William Sound head in the northern and

A climber makes his way across the snow-covered surface of Portlock Glacier, in Kachemak State Park. Unnamed peaks of the Kenai Mountains rise in the background. (Chlaus Lotscher)

western Chugach Mountains. In addition, a number of large valley glaciers descend from the heights of the Chugach Mountains to the Matanuska and Tazlina valleys to the north. Largest of these glaciers are Knik, Matanuska, Nelchina, and Tazlina, with lengths of 25, 28, 25, and 35 miles respectively.

Matanuska Glacier, about 80 miles northeast of Anchorage and visible from the Glenn Highway, drains about 250 square miles of central Chugach Mountains between Mount Marcus Baker and Mount Witherspoon. Matanuska Glacier was investigated in detail in the mid-1950s by John Williams and Oscar Ferrians of the U.S. Geological Survey. Their work showed that little change was occurring in the glacier's position in the 1950s, but that moraines about one-quarter mile in front of the terminus were less than 200 years old. Earlier Holocene advances about 4,000 to 8,000 years ago had left conspicuous moraines one to five miles beyond the present terminus.

Since 1974, research on sediment transport and deposition has been conducted at the western terminus of Matanuska Glacier by scientists from the University of Illinois and the Cold Regions Research and Engineering Laboratory of the U.S. Army Corps of Engineers.

During the summer of 1979, part of Matanuska glacier advanced more than 100 feet in 60 days. The terminus overrode stagnant ice and **ablation moraine** that had resulted from a previous long history of slow melting and retreat. The cause of the 1979 advance is unknown.

Worthington Glacier is a small mountain glacier with an area of about three and one-half square miles located just north of Thompson Pass, about 20 miles north of Valdez. Because its terminus is less than one-half mile from the Richardson Highway, the glacier is easily accessible and is often visited by motorists driving to Valdez. Worthington shows conspicuous evidence of recent retreat as huge fresh lateral and terminal moraines surround the glacier. This is confirmed by photographs dating from 1937. In 1967 a midsummer heat balance study conducted by A. Neil Streten and Gerd Wendler of the University of Alaska confirmed that the observed thermal conditions could easily

Grewingk Glacier cascades down the Kenai Mountains toward Kachemak Bay. Visible from Homer, the glacier covers about 30 square miles and is part of an ice field known as the Grewingk-Yalik glacier complex. (Chlaus Lotscher)

account for the recent retreat and high rate of observed ice melting.

Every winter between 1915 and 1966, except 1963, Knik Glacier advanced against the side of Mount Palmer (6,940 feet) and blocked the meltwater flow of Lake George Glacier, Colony Glacier, and Knik Glacier from entering the Knik River. Each spring, meltwater would accumulate behind the terminus of Knik Glacier and form a large lake, Lake George, which covered an area of as much as 25 square miles. By late June or early July, the lake would overtop its ice dam and begin a flood that would last for about two weeks. During the peak of the *jokulhlaup* as much as 150 million gallons of water would drain from the lake each minute. Since 1966, however, Knik Glacier has failed to advance sufficiently to seal off the flood channel. In late 1980 the river channel between the glacier and Mount Palmer was only about 300 feet wide. All that would be needed for the reformation of Lake George would be a minor re-advance. When or if this will happen is unknown, but an ongoing study by the U.S. Geological Survey may answer these questions.

Harding and Sargent Icefields

Harding and Sargent icefields cover a combined area of more than 2,000 square miles and have glaciers that drain into both Prince William Sound and Cook Inlet. More than 25 glaciers flow from 35-mile by 20-mile Sargent Icefield; largest of these are Chenega and Ellsworth glaciers. Ellsworth extends more than 17 miles and empties into a lake above the head of Day Harbor. Approximately 12-mile-long Chenega covers more than 125 square miles and ends at tidewater in Nassau Fiord of Prince William Sound.

Largest of the Kenai Mountain ice fields is Harding, with dimensions of about 50 miles by 30 miles. Four glaciers with lengths greater than 15 miles descend from its summits. The two largest, Skilak and Tustumena, have lengths of about 20 miles each. McCarty Glacier in McCarty Fiord in the East Arm of Nuka Bay retreated about two miles between 1909 and 1927, an additional 12 miles between 1927 and 1950, and two more miles through 1976. In recent years the position of the glacier's terminus has been relatively stable.

Northwestern Glacier, which extends from the southeastern side of Harding Icefield to tidewater in Northwestern Fiord at the head of Harris Bay, was named by U.S. Grant for Northwestern University in Evanston, Illinois. The glacier has been retreating since the start of the 20th century with a total retreat through the 1970s of about seven miles.

Many glaciers that were tributary to the lower part of Northwestern Glacier at the turn of the century are now small remnant ice patches stranded high above Harris Bay. The largest of these, an unnamed glacier, had a tidewater terminus in 1950, retreated to an elevation of about 1,000 feet by 1964, and by late 1980 had again reached tidewater through a process of avalanching, re-advance, and reconstitution.

The Grewingk-Yalik glacier complex, south of Harding Icefield and visible from Homer across Kachemak Bay, straddles the crest of the central Kenai Mountains, and extends for 26 miles from east to west. Petrof and Yalik glaciers, examined in 1909 by Grant and Higgins, are the major glaciers draining the east side of the complex; Dixon, Portlock, Grewingk, Wosnesenski, and Doroshin glaciers, all visible from Homer, are the major glaciers draining the west side. William H. Dall and G.K. Gilbert investigated Grewingk in 1899. Comparison of recent aerial photographs of the 13-mile-long glacier with their early observations shows that Grewingk, with Yalik and Petrof, have all retreated during the 20th century.

Trail Glacier (also known as Grandview Glacier) has long been a popular spot for cross-country skiers because of its proximity to the Alaska Railroad station at Grandview. This photo, taken about 1948, shows the snow-covered surface of the glacier, crisscrossed with ski trails.
(Sidney Hamilton, Lu Liston Collection)

Talkeetna Mountains

The Talkeetna Mountains, between the Chugach Mountains and Alaska Range, contain a number of valley and cirque glaciers at elevations of 4,000 to 8,000 feet, most of which are unnamed. Glaciers are distributed throughout much of the mountainous uplands, with the largest concentration being on the south side, in the mountains overlooking the headwaters of the Talkeetna River. The area's largest glaciers, including seven longer than five miles, are located here. An unnamed, 10-mile-long glacier which forms the headwaters of Sheep River is the largest. Talkeetna and Chickaloon glaciers, each about eight miles long, are the second and third largest. Their termini are covered with thick ablation moraines, suggesting that both, and perhaps many of the other glaciers in the Talkeetna Mountains, are retreating.

Left — *Two lone skiers join their trails in the fresh snow on Snowbird Glacier, in the Talkeetna Mountains, northeast of Anchorage.*
(Vincent McClelland)
Right — *Two unnamed hanging glaciers tumble from the peaks of the Talkeetna Mountains. The area contains a number of cirque and valley glaciers at elevations of 4,000 to 8,000 feet.*
(John Murk)

Alaska Range

The Alaska Range consists of a number of distinct mountain groups that extend in an arc more than 600 miles from the White River, just east of the Alaska-Canada border, to Iliamna Lake at the head of the Alaska Peninsula. More than 5,000 square miles of the range, which curves inland as the Aleutian Range comes north along Cook Inlet's western shore, are covered by glaciers.

For simplicity of discussion, the Alaska Range will be divided into three segments: eastern Alaska Range; Mount McKinley; and the Mount Gerdine-Mount Spurr or western Alaska Range section.

Eastern Alaska Range

The eastern segment consists of three areas: Mentasta and Nutzotin mountains, east of the Glenn Highway; the Mount Kimball area between the Glenn and Richardson highways; and the Mount Hayes area west of the Richardson Highway.

In the Mentasta and Nutzotin mountains are about 25 small cirque glaciers with a combined area of about five square miles. Only one glacier, three-mile-long Carl, is named. A.H. Brooks and Stephen Capps investigated glaciers in this area briefly just after the turn of the century, but these ice masses have not been studied since.

Glaciers of the Mount Kimball (10,300 feet) area cover about 325 square miles, and most descend from a 50-mile-long interconnected ice field located astride the crest of the eastern Alaska Range. At least 19 valley glaciers that

Gulkana Glacier, located about four miles off the Richardson Highway, was the subject of a joint study by the U.S. Geological Survey and the University of Alaska. The study concentrated on foliation and flow patterns, structure, and ablation. This photo, taken in 1977, shows a U.S.G.S. Gaging Station near the glacier's terminus. (Steve McCutcheon)

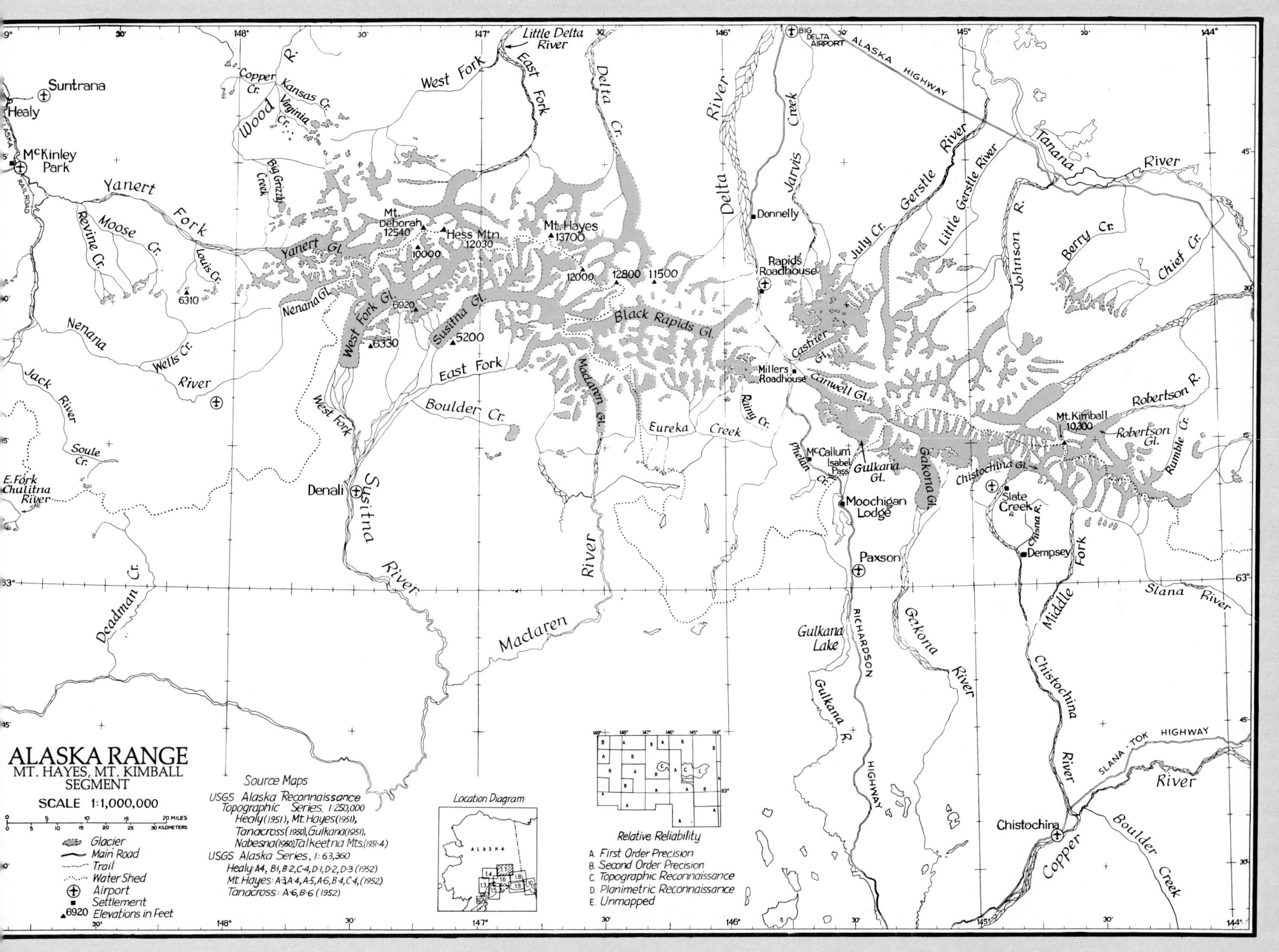
ALASKA RANGE
MT. HAYES, MT. KIMBALL SEGMENT
SCALE 1:1,000,000
20 MILES
30 KILOMETERS
Glacier
Main Road
Trail
Water Shed
Airport
Settlement
6920 Elevations in Feet
Source Maps
USGS Alaska Reconnaissance Topographic Series, 1:250,000
Healy (1951), Mt. Hayes (1951), Tanacross (1950), Gulkana (1951), Nabesna (1950), Talkeetna Mts. (1951-4)
USGS Alaska Series, 1:63,360
Healy: A4, B1, B2, C-4, D-1, D-2, D-3 (1952)
Mt. Hayes: A-3, A-4, A-5, A-6, B-4, C-4, (1952)
Tanacross: A-6, B-6 (1952)
Location Diagram
ALASKA
Relative Reliability
A. First Order Precision
B. Second Order Precision
C. Topographic Reconnaissance
D. Planimetric Reconnaissance
E. Unmapped
Suntrana
Healy
McKinley Park
Yanert Fork
Revine Cr.
Moose Cr.
Louis Cr.
Nenana River
Wells Cr.
Jack River
Soule Cr.
E. Fork Chulitna River
Deadman Cr.
Wood R.
Copper Cr.
Kansas Cr.
Virginia Cr.
Big Grizzly Creek
Yanert Gl.
Nenana Gl.
West Fork Gl.
Susitna Gl.
Mt. Deborah 12540
Hess Mtn. 12030
10000
6920
6310
6330
5200
Mt. Hayes 13700
12000
12800
11500
West Fork
Little Delta River
East Fork
Delta Cr.
Delta River
Black Rapids Gl.
Maclaren Gl.
East Fork
Boulder Cr.
West Fork
Denali
Susitna River
Maclaren River
Eureka Creek
Rainy Cr.
BIG DELTA AIRPORT
ALASKA HIGHWAY
Jarvis Creek
Donnelly
Rapids Roadhouse
Millers Roadhouse
Castner Gl.
Canwell Gl.
Phelan Cr.
McCallum
Isabel Pass
Gulkana Gl.
Gakona Gl.
Moochigan Lodge
Paxson
Gulkana Lake
Gulkana R.
RICHARDSON HIGHWAY
Gakona River
Gerstle River
Little Gerstle River
July Cr.
Johnson R.
Tanana River
Berry Cr.
Chief Cr.
Robertson R.
Mt. Kimball 10,300
Robertson Gl.
Rumble Cr.
Chistochina Gl.
Slate Creek
Chisna R.
Dempsey
Middle Fork
Slana River
Chistochina River
SLANA-TOK HIGHWAY
Copper River
Boulder Creek
Chistochina
148°
147°
146°
145°
144°
63°

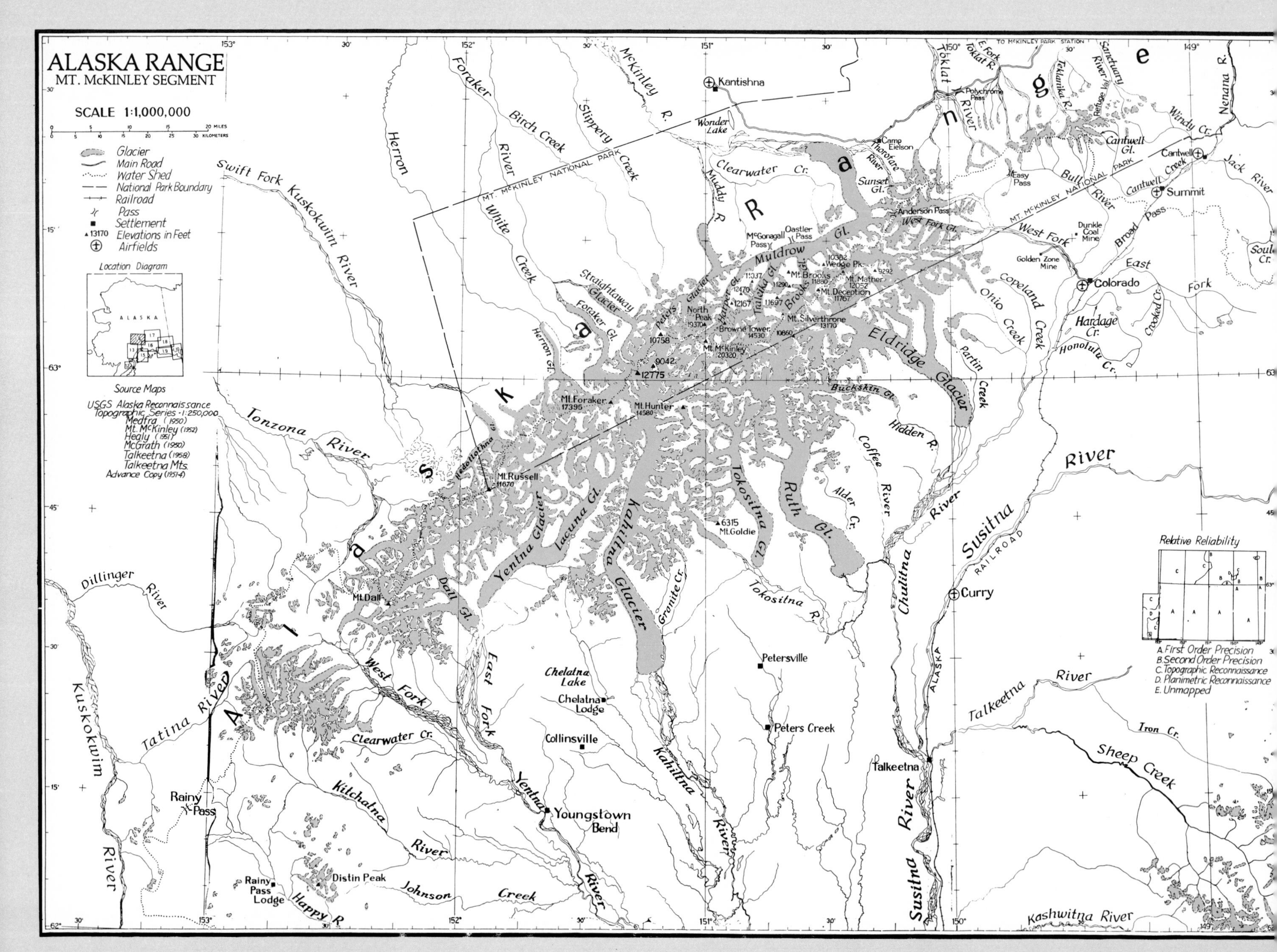
ALASKA RANGE
MT. McKINLEY SEGMENT
SCALE 1:1,000,000
MILES
KILOMETERS
Glacier
Main Road
Water Shed
National Park Boundary
Railroad
Pass
Settlement
13170 Elevations in Feet
Airfields
Location Diagram
ALASKA
Source Maps
USGS Alaska Reconnaissance Topographic Series -1:250,000
Medfra (1950)
Mt. McKinley (1952)
Healy (1951)
McGrath (1950)
Talkeetna (1958)
Talkeetna Mts.
Advance Copy (1951)
Relative Reliability
A. First Order Precision
B. Second Order Precision
C. Topographic Reconnaissance
D. Planimetric Reconnaissance
E. Unmapped
A l a s k a R a n g e
TO McKINLEY PARK STATION
MT. McKINLEY NATIONAL PARK
Kantishna
Wonder Lake
Camp Eielson
Polychrome Pass
Sunset Gl.
Easy Pass
Cantwell Gl.
Cantwell
Summit
Anderson Pass
West Fork Gl.
McGonagall Pass
Oastler Pass
Muldrow Gl.
Wedge Pk.
Mt. Brooks
Mt. Mather
Mt. Deception
Mt. Silverthrone
North Peak
Browne Tower
Mt. McKinley
20320
Mt. Foraker
17395
Mt. Hunter
14580
Mt. Russell
11670
Mt. Goldie
Mt. Dall
Distin Peak
Foraker River
Birch Creek
Slippery Creek
McKinley R.
Herron River
White Creek
Swift Fork Kuskokwim River
Straightaway Glacier
Foraker Gl.
Peters Glacier
Herron Gl.
Clearwater Cr.
Muddy R.
Thorofare River
Toklat River
E. Fork Toklat R.
Teklanika R.
Sanctuary River
Nenana R.
Windy Cr.
Jack River
Bull River
Cantwell Creek
West Fork
Dunkle Coal Mine
Golden Zone Mine
Broad Pass
East Fork
Colorado
Copeland Creek
Ohio Creek
Hardage Cr.
Crooked Cr.
Honolulu Cr.
Partin Creek
Eldridge Glacier
Buckskin Gl.
Hidden R.
Coffee River
Alder Cr.
Ruth Gl.
Tokositna Gl.
Tokositna R.
Kahiltna Glacier
Lacuna Gl.
Yentna Glacier
Dall Gl.
Granite Cr.
Tonzona River
Dillinger River
Kuskokwim River
Tatina River
West Fork
Clearwater Cr.
Kitchatna River
Rainy Pass
Rainy Pass Lodge
Happy R.
Johnson Creek
Yentna River
Youngstown Bend
East Fork
Chelatna Lake
Chelatna Lodge
Collinsville
Kahiltna River
Petersville
Peters Creek
Talkeetna
Susitna River
Chulitna River
Curry
ALASKA RAILROAD
Talkeetna River
Iron Cr.
Sheep Creek
Kashwitna River

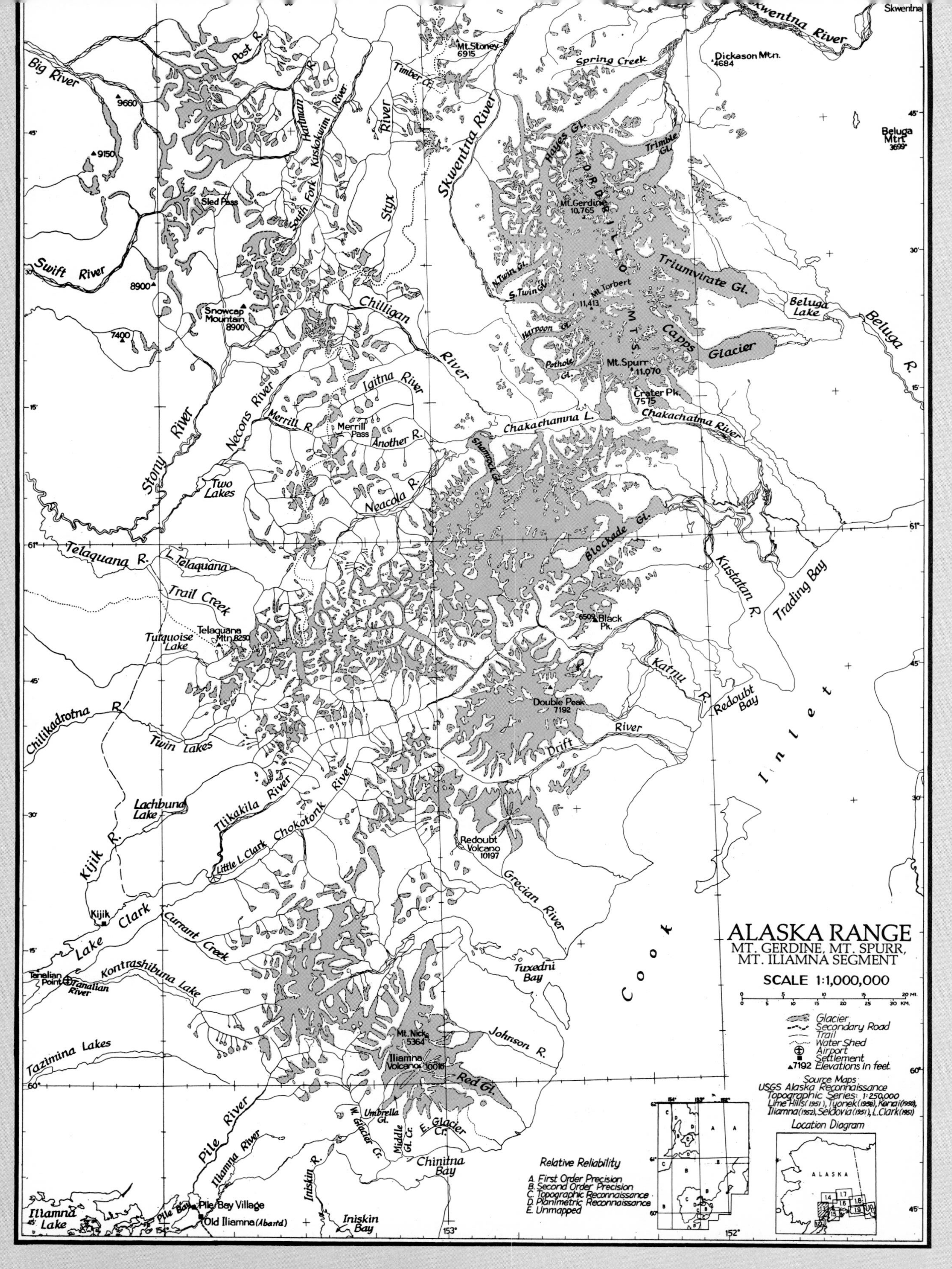
ALASKA RANGE
MT. GERDINE, MT. SPURR,
MT. ILIAMNA SEGMENT
SCALE 1:1,000,000
Glacier
Secondary Road
Trail
Water Shed
Airport
Settlement
▲7192 Elevations in feet
Source Maps
USGS Alaska Reconnaissance
Topographic Series: 1:250,000
Lime Hills (1951), Tyonek (1958), Kenai (1950),
Iliamna (1952), Seldovia (1951), L. Clark (1951)
Location Diagram
ALASKA
Relative Reliability
A. First Order Precision
B. Second Order Precision
C. Topographic Reconnaissance
D. Planimetric Reconnaissance
E. Unmapped
Skwentna
Skwentna River
Dickason Mtn. 4684
Beluga Mtn. 3699
Spring Creek
Mt. Stoney 6915
Post R.
Big River
9660
9150
Hartman R.
South Fork Kuskokwim River
Timber Cr.
Styx River
Skwentna River
Hayes Gl.
Trimble Gl.
T O R D R I L L O M T S.
Mt. Gerdine 10,765
Sled Pass
Swift River
8900
Snowcap Mountain 8900
7400
N. Twin Gl.
S. Twin Gl.
Triumvirate Gl.
Mt. Torbert 11,413
Beluga Lake
Beluga R.
Chilligan River
Harpoon Gl.
Capps Glacier
Pothole Gl.
Mt. Spurr 11,070
Crater Pk. 7575
Igitna River
Necons River
Merrill R.
Merrill Pass
Another R.
Chakachamna L.
Chakachatna River
Stony River
Two Lakes
Shamrock Gl.
Neacola R.
Blockade Gl.
Telaquana R.
L. Telaquana
Trail Creek
Kustatan R.
Trading Bay
6509 Black Pk.
Telaquana Mtn. 8250
Turquoise Lake
Katnu R.
Double Peak 7192
Redoubt Bay
Chilikadrotna R.
Twin Lakes
Drift River
Cook Inlet
Lachbuna Lake
Tlikakila River
Chokotonk River
Kijik R.
Little L. Clark
Redoubt Volcano 10197
Grecian River
Kijik
Lake Clark
Currant Creek
Tanalian Point
Tanalian River
Kontrashibuna Lake
Tuxedni Bay
Tazimina Lakes
Mt. Nick 5364
Iliamna Volcano 10016
Johnson R.
Red Gl.
Pile River
Iliamna River
Umbrella Gl.
W. Glacier Cr.
Middle Gl. Cr.
E. Glacier Cr.
Chinitna Bay
Iniskin R.
Iniskin Bay
Iliamna Lake
Pile Bay
Pile Bay Village
Old Iliamna (Aband.)

Left — *Although presently retreating, Black Rapids Glacier was nicknamed the "Galloping Glacier" because of its remarkable surge during the winter of 1936-37. At that time, the glacier advanced more than three miles in only six months, reaching a position that threatened the Richardson Highway. This, fortunately, was its maximum position.* (Steve McCutcheon)
Right — *The surface of Mount McKinley (20,320 feet), sculptured by countless years of glacial action, gives rise to many present-day glaciers. The lake in the foreground — a kettle — is a remnant of an ancient glacier.* (John Johnson)

head along the crest have lengths of five miles or more. The largest glacier is 75-square-mile, 18-mile-long Johnson Glacier which forms the headwaters of Johnson River. No long-term history of observation exists for most of these glaciers with the exception of Castner, Canwell, and Gulkana.

Canwell Glacier, site of a camping and skiing festival held every spring, was first investigated in 1902 by Walter C. Mendenhall and Thomas G. Gerdine. The 15-mile-long glacier and its neighbor, Castner Glacier, were both studied in the 1950s by Troy Pewe of the University of Alaska and the U.S. Geological Survey. Pewe found that both glaciers had advanced about one mile at the start of the Little Ice Age, and that Canwell Glacier had a second smaller advance within the past 200 years. Canwell's advance was then followed by a retreat of more than a mile. In the early part of this century, between 1902 and 1941, Canwell Glacier advanced once again. Photographic comparisons suggest that this advance was also about one mile.

In the early 1960s Gulkana Glacier, about four miles east of the Richardson Highway, was the subject of intensive study by the University of Alaska and the U.S. Geological Survey to characterize the geophysical parameters of the glacier, foliation patterns, structure, flow patterns, and ablation. Pewe, Ned Ostenso, and Paul Sellmen determined the configuration of Gulkana Glacier's valley by a gravimeter (a portable instrument to measure gravity) survey of the ice surface in 1961. They discovered that Gulkana's valley consists of two parallel bedrock channels separated by a medial ridge. Ice in the eastern channel was about 750 feet thick, while in the shallower western valley it was only about 430 feet thick. Gulkana Glacier was photographed by Fred Howard Moffit in 1910. Photographs taken in 1960, from Moffit's photo site, and photographs made in subsequent years show that the terminus has retreated several miles. Beginning in 1966, annual mass balance studies of Gulkana Glacier have been conducted.

Ned Ostenso studied Jarvis Glacier, just north of Gulkana Glacier, using gravimetric techniques in 1955. His gravity survey showed that Jarvis Glacier lies in a deep, U-shaped valley with ice thicknesses of about 1,050 feet.

The Mount Hayes section, which includes

Mount Hayes (13,832 feet), Mount Moffitt (13,020 feet), and Mount Deborah (12,339 feet) has a glacier-covered area of approximately 400 square miles containing 15 glaciers with lengths greater than five miles. The longest are Black Rapids and Susitna glaciers, each about 25 miles long. Like the majority of glaciers in the eastern Alaska Range, most of the ice masses in this area are also retreating.

Some glaciers, however, like Black Rapids, have interrupted their retreats with catastrophic surges. Between 1912 and 1936 Black Rapids Glacier retreated more than three miles. In September or October 1936 the glacier began to surge. During the winter of 1936-1937 Black Rapids Glacier advanced more than three miles in only six months. By February 1937 the glacier had reached a position that threatened the Richardson Highway, adjacent to the Delta River. Here, fortunately, Black Rapids reached its maximum advance. Retreat and downwasting began and have lasted to the present. During the period of maximum surge, Black Rapids advanced as much as 115 feet per day, and consequently was nicknamed the "Galloping Glacier."

A historical marker placed at the edge of the Richardson Highway overlooking Black Rapids Glacier identifies the terminal moraine from the 1937 surge. It is interesting to note that the sign itself stands on the moraine of an even earlier and larger surge. In late 1980 Black Rapids Glacier was being closely watched by the U.S. Geological Survey because it appears to be building up to yet another surge.

Susitna and Yanert glaciers have also surged within the past 40 years. Yanert Glacier, west of Mount Deborah, advanced three miles in 1942, while in 1952 or 1953 Susitna Glacier surged a similar distance.

Mount McKinley Area

From the Mount McKinley segment of the Alaska Range, about 175 miles long and almost 50 miles wide, rise some of the highest mountains in North America: Mount McKinley (20,320 feet), and Mount Foraker (17,400 feet), third highest peak in Alaska, are the tallest. More than 20 glaciers with lengths greater than five miles descend from a broad *névé* in this, the highest section of the range. Six glaciers have lengths of 25 miles or more: Kahiltna (43 miles), Muldrow (40 miles), Ruth (36 miles), Yentna-Lacuna (32 miles), Eldridge (30 miles), and Tokositna (25 miles). Both Muldrow and Kahiltna are used as access routes to Mount McKinley by climbers attempting to reach the continent's highest peak.

Except for Ruth Glacier, which reaches the Chulitna Valley, the glaciers on the south side of the range all terminate in U-shaped valleys that extend up to 30 miles beyond the present ice fronts. On the north side of the range the large glaciers extend up to five miles beyond

Frances Randall: Queen of Kahiltna Glacier

By Amy Bermar, Anchorage Daily News

Kahiltna Glacier has an unofficial queen, an unpretentious airport, and more than 135 square miles of snow and ice.

It also has a steady flow of avalanches, temperatures that can drop 70° in an hour, and a radio station.

Kahiltna's queen is Frances Randall, manager of the base camp that is nine miles from the south side of Mount McKinley, and the starting and finishing point for more

Clockwise from left — *This avalanche came spilling down Mount Hunter (14,573 feet) one sunny lunchtime. Base camp residents can hear — or see — as many as a dozen avalanches in a single day, especially in warm weather following a storm.* (Amy Bermar, *Anchorage Daily News*) ►*Base camp manager Frances Randall watches a plane preparing to land. Behind are the plastic sleds used by climbers to carry their gear.* (Amy Bermar, *Anchorage Daily News*) ►*Lowell Thomas brings his Helio Courier in for a landing. The black spots are plastic garbage bags draped over wire rings, the only markings on the stomped-out landing strip. Glacier pilots call camp manager Frances Randall for a weather report before even taking off, since the skies may be clear in Talkeetna, but there could be a whiteout around McKinley.* (Dave Blanchett) ►*Steve Reidel (left) and Bo McFadden arrange their gear during the days before their climb. The camp hosted about 500 climbers from around the world during the summer of 1980. As many as 50 may be in camp at once, some of whom have completed their climbs and are waiting for the weather to clear so they can fly home.* (Dave Blanchett)

than 500 climbers who try to reach the 20,320-foot summit each summer.

Randall, who doubles as a symphony violinist and computer analyst at the University of Alaska, Fairbanks, during the winter, has managed the camp since 1974. Her home — a canvas tent on floorboards, reconstructed each summer — comes complete with a mattress, stacks of books, and the two radios that are her only connection to the outside. She wears dark sunglasses as protection against the almost constant glare, and black rubber boots on warm days when the snow is wet and heavy.

Because Kahiltna is a moving body of ice, there are constant moans and groans as the ice shifts, forming deep crevasses that more than one unsuspecting climber has fallen into. Avalanches are another occupational hazard of living on a glacier, as they come rumbling down one of the dozens of slopes surrounding the glacial bowl in which the camp is located.

But despite the dangers, Randall is content with her summer lodging.

"I have all the comforts of home here," she says with an expansive wave around her tent, "except hot water."

A backpacker closely inspects a large granite boulder — known as a glacial erratic —transported and left high on the tundra by an ancient glacier. This photo, taken in Denali National Park and Preserve, includes a view of Mount Hayes (13,832 feet) and Mount Deborah (12,339 feet) on the left. (John Johnson)

the mountain front and terminate on a low-relief plateau within Denali National Park and Preserve.

A study published in 1980 by N.W. Ten Brink and D.R. Ritter documented the late Pleistocene and early Holocene history of the northcentral Alaska Range. Beginning about 25,000 years ago and ending by about 9,500 years ago, a major glacial event, the McKinley Park Glaciation, influenced most of the Mount McKinley region. Four major glacial advances occurred: about 20,000 years ago, 15,000 to 13,500 years ago, 12,800 to 11,800 years ago, and 10,500 to 9,500 years ago. Each period of advance was followed by a warming trend that resulted in rapid ice melting and retreat. Following the 9,500-year-old ice advance, warm climatic conditions persisted until about 6,000 years ago. Ten Brink and Ritter point out that the first sites of human habitation in the Alaska Range date from the interval between the last two ice advances. They suggest that such sites may be found along the large braided streams that drained the shrinking glaciers of the early Holocene. These streams were environments where grasses could grow and attract large grazing animals on which early Alaska man preyed.

Muldrow Glacier, covering about 200 square miles, flows northeast from high on Mount McKinley's slopes to a moraine-covered terminus near Eielson Visitor Center

Left — *Large kettle and ice cave found in the 150-year-old moraine of Muldrow Glacier, which heads high on the north side of Mount McKinley. The soil covering the stagnant ice is more than four feet thick in some places.*
Below — *The Kahiltna River snakes through a lush valley at the terminus of Kahiltna Glacier. The glacier, which flows from an ice field between Mount Foraker and Mount Hunter, is used as an access route by climbers attempting to reach the summit of Mount McKinley.* (Steve McCutcheon)

in Denali National Park and Preserve. Two named tributary glaciers, Brooks and Traleika, furnish much of Muldrow's ice. Beginning in May of 1956 and continuing through the summer of 1957, Muldrow Glacier surged a distance of about four miles. The maximum observed velocity during the peak of movement was about 1,150 feet per day, or almost 10 inches per minute. For almost a decade prior to the surge, a wave of thickening ice moved down the upper part of the glacier at a rate of about two and one-fourth feet per day. A.E. (Will) Harrison, who investigated the 1956-1957 surge, reported that an analysis of moraine patterns on the glacier's surface suggest that at least four prior surges occurred within the past 200 years, with the most recent pre-1956 surge occurring between 1906 and 1912.

Muldrow Glacier has deposited three terminal moraine complexes during the past 300 years. The largest moraine, clearly visible from Eielson Visitor Center and the park road, was formed by a 16th and 17th century advance that may represent the maximum ice position of Muldrow Glacier in Neoglacial time. Ice retreat followed, but about 150 years ago was interrupted by a re-advance that deposited a second set of moraines, one to three miles behind the 17th century moraine. The 1957 Muldrow Glacier surge overrode much of the second set of moraines, and established a terminal position within three miles of the park road. Today ablation moraine covers the surface of much of the lower five miles of Muldrow Glacier.

Gerdine-Spurr Area

The Tordrillo Mountains, which include Mount Gerdine (11,258 feet), and Mount Spurr (11,100 feet), contain the largest glaciers in this segment. Hayes, Trimble, Triumvirate, and Capps glaciers, all of which drain eastward, reach lengths of 15 to 25 miles. An eruption of Mount Spurr in 1953 dusted glaciers with ash as far away as the Chugach Mountains.

To the west of the Tordrillo Mountains, a number of smaller glaciers lie on the summits and flanks of the Revelation Mountains, and on a number of peaks between the Skwentna, Stoney, and Styx rivers. These are generally valley glaciers with lengths of less than five miles and widths of a mile or less. Revelation Glacier is the largest of the entire group with a length of almost 10 miles.

Many small unnamed cirque glaciers perch among the peaks of Rainy Pass, in the Teocalli Mountains 96 miles southeast of McGrath. The mountains were named in 1898 by Josiah Spurr, a geologist with the U.S. Geological Survey, who thought they looked like Aztec temples. (Betty Johannsen)

Aleutian Range

The Aleutian Range extends for 600 miles, in a northeast-southwest direction, along the Alaska Peninsula and into the Aleutian Islands. At least six areas of the range support glaciers. Other areas might also contain active glaciers, but there is a general absence of glaciological (ice) and glacial geological (rocks and features created by ice) information about the Aleutian Range in scientific literature. Areas of the range that presently have active glaciers include: the Chigmit Mountains; Katmai National Park and Preserve; the Mount Kialagvik-Icy Peak-Mount Chiginagak area, southwest of Wide Bay; Aniakchak Crater, north of Kujulik Bay; the Mount Veniaminof area, northwest of Stepovak Bay; and the area around Pavlof Volcano, west of Pavlof Bay. The glaciated area of the Aleutian Range exceeds 1,000 square miles.

The Chigmit Mountains at the northern extreme of the Aleutian Range claim Redoubt Volcano (10,197 feet) and Iliamna Volcano (10,016 feet), both glacier-covered, active volcanoes, as their highest peaks. In addition to these peaks, the northern Chigmits contain more than 200 square miles of *névé* and valley glaciers. One glacier, Blockade, divides with one terminus damming Blockade Lake and the other flowing into McArthur River, more than 10 miles to the southwest. Another ice mass, Shamrock Glacier, flows north for 16 miles and has its terminus pushing across Chakachamna Lake. At its maximum position, the glacier separated Kenibuna Lake from Chakachamna and almost created an ice dam more than three miles long.

More than a dozen glaciers, all unnamed, descend from the heights of Redoubt Volcano. In January 1966, when the mountain last erupted, much melting of the summit glaciers took place, followed by outburst flash flooding on Drift River. Floods in 1966 covered the lower two and one-half miles of the northernmost glacier that descends from Redoubt Volcano with volcanic debris to a maximum depth of up to six and one-half feet. At present the debris-covered terminus is less than 300 feet from the far wall of the Drift River Valley.

Since 1977 the University of Alaska has been investigating glaciers at the summit and on the north flank of Redoubt Volcano to determine what the role of volcanic action has been and will be on changes in ice conditions. In 1979 Peter MacKeith, Carl Benson, and Jurgen Kienle reported that the effect of the 1966 eruption had not yet reached the lower part of the glacier. They observed ice thickening and rapid motion about one and one-half miles above the terminus. Similar observations were made on other parts of the glacier. These investigations suggest that this minor advance, possibly caused by the insulating effects of the flood debris, could dam the Drift River and lead to formation of an ice-dammed lake. The major tanker loading facility for petroleum drilled in Cook Inlet is at the mouth of the Drift River. An outburst flood from the glacier-dammed lake could create a potentially catastrophic situation for the tanker terminal.

Iliamna Volcano, like Redoubt Volcano, has more than a dozen large glaciers that descend from its summit. Iliamna's rivers of ice are larger, and cover about four times the area of Redoubt's. Tuxedni Glacier, largest on Iliamna with a length of 16 miles, extends from the volcano's summit to sea level tide

Iliamna Volcano (10,016 feet), in the Chigmit Mountains, is an active volcano with more than a dozen large glaciers flowing from its slopes. This photo shows the north side of Iliamna, with Tuxedni Glacier at the right and the Tuxedni River winding around the volcano's base.
(John and Margaret Ibbotson)

ALEUTIAN RANGE

For continuation of the Aleutian Range see page 123

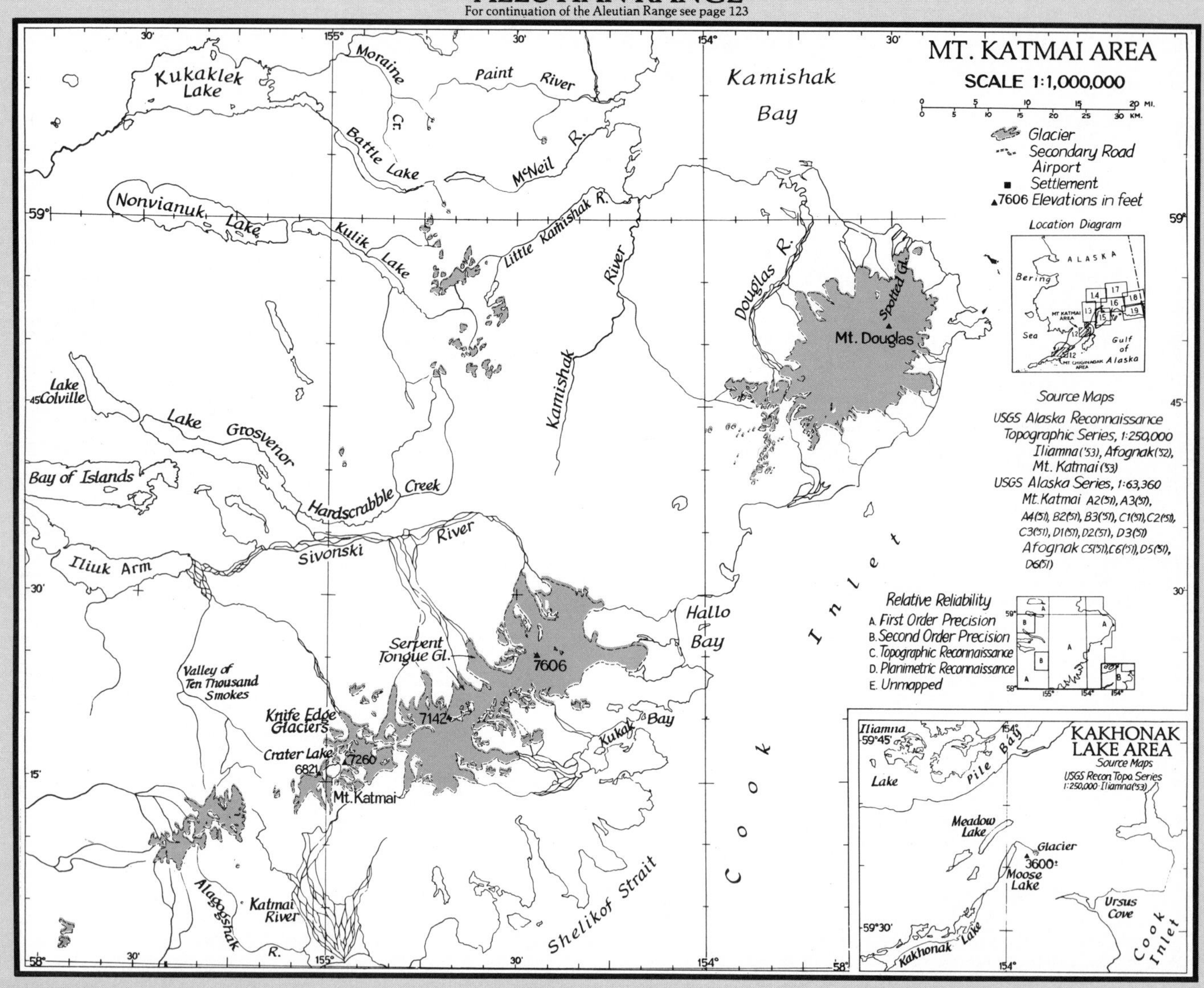

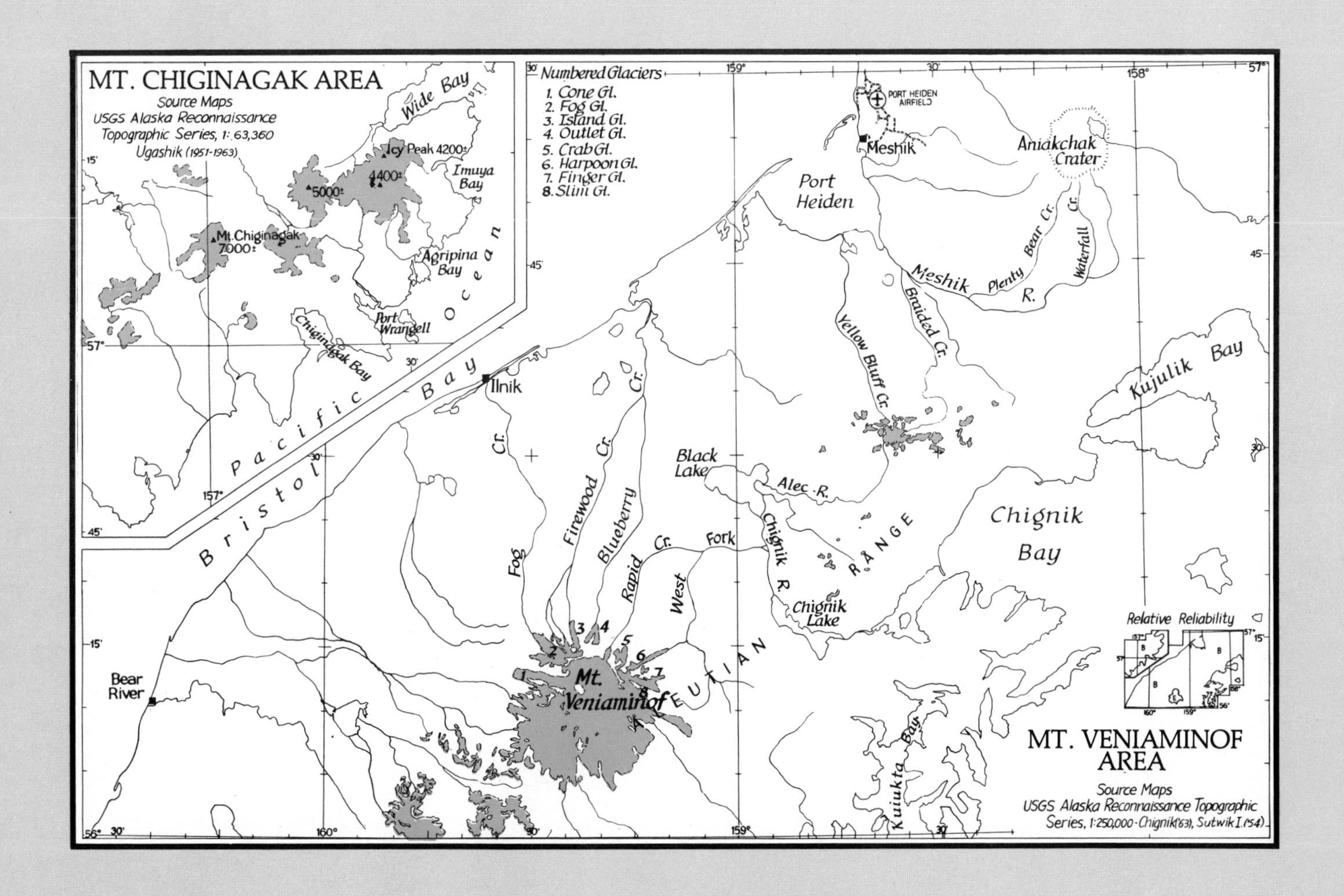
MT. CHIGINAGAK AREA
Source Maps
USGS Alaska Reconnaissance
Topographic Series, 1: 63,360
Ugashik (1951-1963)
Wide Bay
Icy Peak 4200±
4400±
5000±
Imuya Bay
Mt. Chiginagak 7000±
Agripina Bay
Port Wrangell
Chiginagak Bay
Pacific Ocean
Numbered Glaciers
1. Cone Gl.
2. Fog Gl.
3. Island Gl.
4. Outlet Gl.
5. Crab Gl.
6. Harpoon Gl.
7. Finger Gl.
8. Slim Gl.
PORT HEIDEN AIRFIELD
Meshik
Aniakchak Crater
Port Heiden
Meshik
Plenty Bear Cr.
Waterfall Cr.
R.
Braided Cr.
Yellow Bluff Cr.
Kujulik Bay
Bristol Bay
Ilnik
Cr.
Firewood Cr.
Blueberry
Black Lake
Alec R.
Chignik Bay
Fog
Rapid Cr.
West Fork
Chignik R.
RANGE
Chignik Lake
ALEUTIAN
Mt. Veniaminof
Bear River
Kuiukta Bay
Relative Reliability
MT. VENIAMINOF AREA
Source Maps
USGS Alaska Reconnaissance Topographic
Series, 1:250,000 - Chignik('63), Sutwik I.('54)

Left — *Smoke billows from the summit crater of Mount Veniaminof (8,225 feet), an active volcano in the Aleutian Range. The south side of the volcano's summit is covered by a large, unnamed, semi-circular glacier, which covers an area of about 60 square miles. Veniaminof last erupted in 1944.* (Lee Miller)
Right — *The rugged Aghileen Pinnacles (4,800 feet), located 24 miles northeast of Cold Bay, are the site of many small cirque glaciers. Pavlof Volcano (8,905 feet) rises in the background.* (John Sarvis)

flats in Tuxedni Bay. Red Glacier, named for the iron-rich, red-colored moraines that cover its lower three miles, is also more than 10 miles long and terminates at an elevation of about 250 feet at the head of the Red River. Other major glaciers that drain Iliamna's flanks are Lateral, Johnson, and Umbrella.

Literally hundreds of glaciers, 20 of which have lengths of five miles or more, are located within Katmai National Park and Preserve. One unique feature of the park and preserve are two small glaciers that have formed in Mount Katmai's caldera, which was created by the 1912 eruption of Novarupta, a little more than seven miles west of the mountain. Prior to this eruption, Mount Katmai's summit was similar to that of Redoubt and Iliamna volcanoes, being completely encircled by active glaciers. As a result of the eruption about 2,000 feet of Katmai's summit, including snowfields and glaciers, disappeared, leaving many beheaded glaciers. More than one and one-half cubic miles of volcanic material was ejected by the 1912 eruption, much of which fell as ash throughout the northern hemisphere. Glaciers on Mount Katmai and on adjacent peaks are covered by ash and pumice deposits that exceed several feet in thickness in a number of places.

Glaciers at the summits and on the slopes of Mount Chiginagak, Mount Kialagvik, and Aniakchak Crater each cover areas of about 10 square miles or less and are restricted to elevations above 3,000 feet. Only one glacier, an unnamed ice mass about four miles long, descends from Mount Kialagvik to near the 1,000-foot level. Its terminus forms the headwaters of Dog Salmon River.

Icy Peak (4,500 feet) supports more than a dozen small, unnamed glaciers that flow down from a five-mile-long upland to elevations below 1,000 feet. One five-mile-long river of ice reaches the headwaters of Glacier Creek, less than 600 feet above sea level.

Topographic maps of Mount Veniaminof (8,225 feet) show the summit and southern slopes blanketed by a large glacier covering more than 60 square miles. Eight named valley glaciers — Cone, Fog, Island, Outlet, Crab, Harpoon, Finger, and Slim — with a combined area of about four square miles descend from the summit region and a number of others occur to the southwest, east, and northeast. Slim Glacier's terminus descends to an elevation of less than 1,000 feet at a point more than 11 miles from the summit crater.

The more than 50 square miles of glaciated area near Pavlof Volcano have the southernmost glaciers in the continental Aleutian Range. Much of the upper slopes of Pavlof Volcano, Litte Pavlof Volcano, Double Crater, Pavlof Sister Volcano, Mount Hague, and Mount Emmons are covered by snowfields and ice at elevations above 1,000 feet. None of these glaciers have been studied in recent years.

Aleutian Islands

The Aleutian Islands, which contain more than 25 active volcanoes, extend 1,100 miles from the Alaska Peninsula to Attu Island. Ice masses are present on at least eight islands in the chain: Unimak, Unalaska, Umnak, Atka, Great Sitkin, Tanaga, Gareloi, and Kiska. Without exception, Aleutian Islands glaciers descend from near-summit levels of active and dormant volcanoes.

On Unimak Island, easternmost and largest of the Aleutian Islands, glaciers are concentrated on the Shishaldin Volcano-Isanotski Peaks-Roundtop Mountain complex in the southeast, and on Pogromni Volcano in the southwest. Summits of the Shishaldin (9,372 feet) complex have a continuous ice field 22 miles long with an area of more than 60 square miles. Pogromni Volcano supports another group of small glaciers covering about 10 square miles.

Makushin Volcano (6,680 feet) on Unalaska Island is topped by an icecap of unknown dimensions. The most recent U.S. Geological Survey topographic map of the island does not even show a glacier cover on Makushin Volcano. Descriptions of glaciers on Unalaska can be found in scientific literature back to the turn of the century. Robert Sharp suggests that Unalaska Island has the largest ice cover of any island in the chain. A 1961 U.S. Geological Survey report indicates that Makushin Volcano is covered by a small, crater-filling glacier which nurtures several small valley glaciers descending down the mountain's flanks. The Shaler Mountains, in the south-central part of Unalaska Island, also support a few small glaciers.

Umnak Island has many valley glaciers that descend from Mounts Vsevidof and Recheshnoi, and at least one small summit glacier in Okmok Caldera. A 1945 eruption of

Right — *The upper slopes of Mount Vsevidof, on Umnak Island, are the source of two unnamed valley glaciers. Without exception, the glaciers of the Aleutian Islands descend from near-summit levels of active and dormant volcanoes.* (Thomas P. Miller, U.S.G.S.)

Opposite — *The largest concentration of glaciers on Unimak Island is found in a continuous, 60-square-mile ice field which stretches across the summits of Shishaldin Volcano, Isanotski Peaks, and Roundtop Mountain. This photo shows Isanotski Peaks, classic examples of a cone altered by volcanic activity, with Shishaldin steaming in the background.* (John Sarvis)

Okmok Caldera melted a large area of the caldera's glacier. Mount Vsevidof's summit is the source of two valley glaciers while Mount Recheshnoi supports at least seven.

On Atka Island two volcanoes have summit glaciers. Korovin Volcano (4,852 feet) and Mount Kliuchef (3,000 feet) have a total glacier-covered area of less than 20 square miles.

Great Sitkin Island, northeast of Adak, has at least five small glaciers that surround the summit of Great Sitkin Volcano (5,740 feet). Largest of these glaciers, all of which descend from the island's ice-ringed crater, is less than two miles long.

A U.S. Army Map Service map of Tanaga Island, done in 1957, shows a few glaciers about one mile in length on Tanaga Volcano (5,925 feet) and on an unnamed volcano to the east. The most recent U.S. Geological Survey map of the area does not show glaciers on Tanaga Island.

Gareloi and Kiska islands both possess small, glacier-covered areas. Gareloi Volcano (5,160 feet) has two small glaciers on its cone while Kiska Volcano (3,996 feet) has a very small residual glacier. In 1947 K.A. Henderson and William L. Putnam described it as being a "small, decadent, dirty piece of ice, but it still merited the name glacier."

THE ALEUTIAN ISLANDS

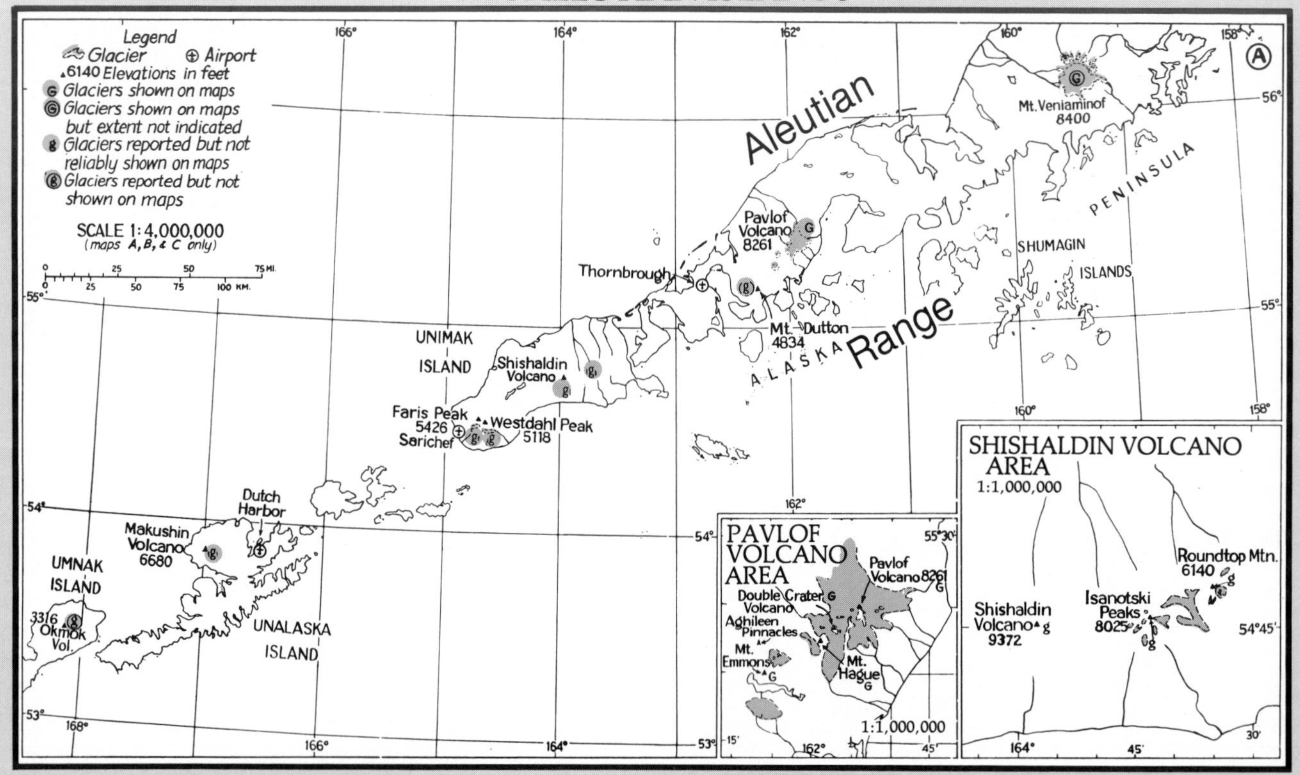

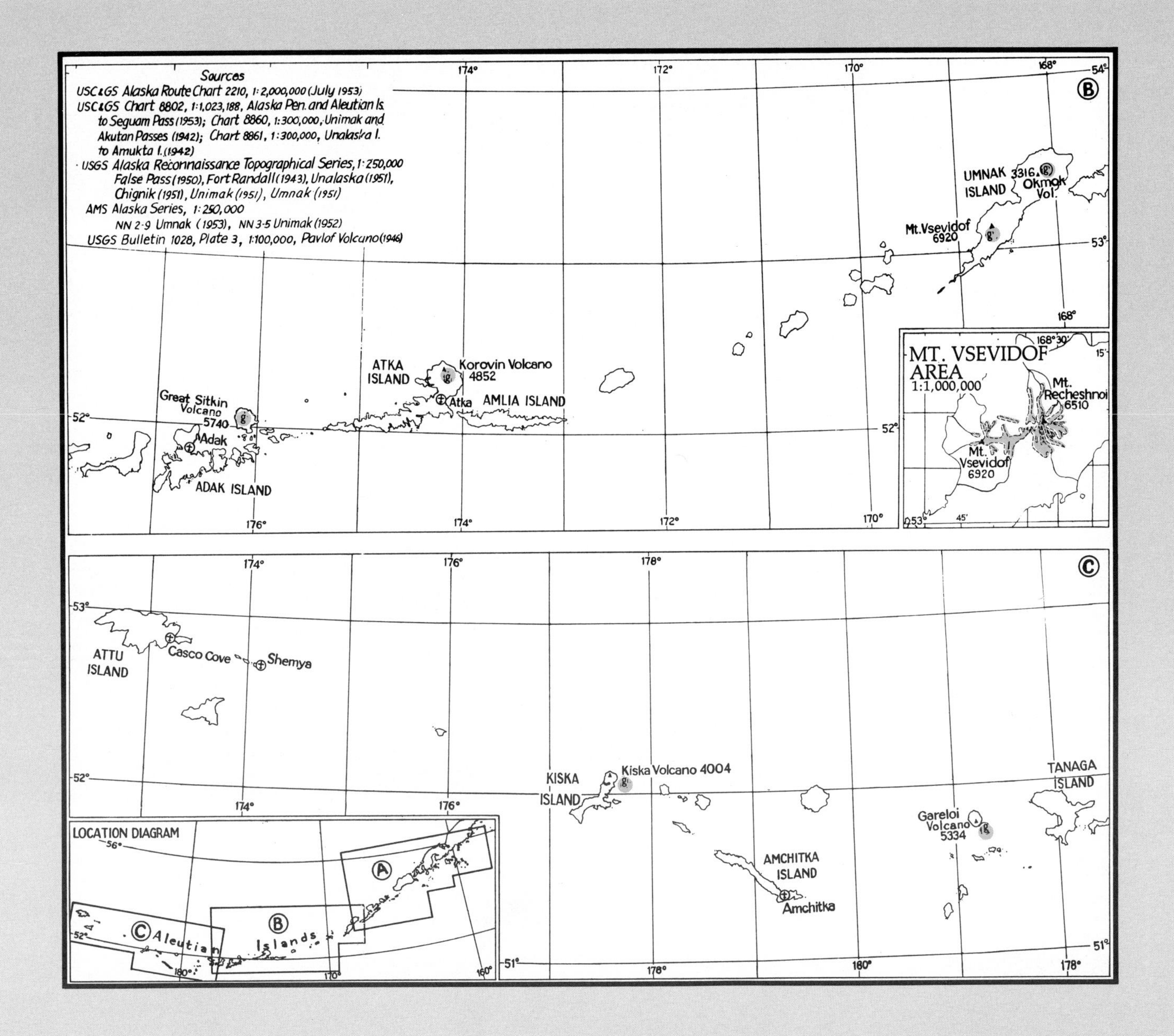
Sources
USC&GS Alaska Route Chart 2210, 1:2,000,000 (July 1953)
USC&GS Chart 8802, 1:1,023,188, Alaska Pen. and Aleutian Is. to Seguam Pass (1953); Chart 8860, 1:300,000, Unimak and Akutan Passes (1942); Chart 8861, 1:300,000, Unalaska I. to Amukta I. (1942)
USGS Alaska Reconnaissance Topographical Series, 1:250,000 False Pass (1950), Fort Randall (1943), Unalaska (1951), Chignik (1951), Unimak (1951), Umnak (1951)
AMS Alaska Series, 1:250,000 NN 2-9 Umnak (1953), NN 3-5 Unimak (1952)
USGS Bulletin 1028, Plate 3, 1:100,000, Pavlof Volcano (1946)
B
UMNAK ISLAND
3316
Okmok Vol.
Mt. Vsevidof 6920
ATKA ISLAND
Korovin Volcano 4852
Atka
AMLIA ISLAND
Great Sitkin Volcano 5740
Adak
ADAK ISLAND
MT. VSEVIDOF AREA
1:1,000,000
Mt. Recheshnoi 6510
Mt. Vsevidof 6920
C
ATTU ISLAND
Casco Cove
Shemya
KISKA ISLAND
Kiska Volcano 4004
TANAGA ISLAND
Gareloi Volcano 5334
AMCHITKA ISLAND
Amchitka
LOCATION DIAGRAM
A
B
C
Aleutian Islands

Kodiak Island

Kodiak Island lies in the Gulf of Alaska, south of Cook Inlet and east of Shelikof Strait. Much, if not all, of the 100- by 60-mile island was covered by a large ice field during the Pleistocene. Today 40 unnamed cirque glaciers survive in a narrow upland region in the center of the island between Koniag Peak (4,470 feet) and Mount Glottof (4,405 feet).

Kodiak's two largest glaciers are about two miles long. In all, less than 20 square miles of Kodiak Island is presently glaciated.

Left — *A small cirque glacier cascades down a slope on Kodiak Island. During the Pleistocene, much, if not all, of the island was covered by a large ice field.* (Mike Vivion, U.S.F.&W.S.)
Above — *Today approximately 40 small cirque glaciers exist in a narrow region in the center of Kodiak Island, between Koniag Peak (4,470 feet) and Mount Glottof (4,405 feet). Note the glacially eroded topography of the island — evidence of relatively recent occupancy by glaciers.*

Kilbuck and Wood River Mountains

The Kilbuck Mountains, 80 miles east of Bethel, and the Wood River Mountains, 60 miles northeast of Dillingham, were glacier-covered during the Pleistocene and have supported glaciers at times during the Holocene. Today at least five alpine and valley glaciers still exist in the Wood River Mountains. Existence of present-day glaciers in the Kilbuck Mountains, however, is not confirmed.

At least four glaciers, all less than three miles long, are clustered on the north slopes of 5,026-foot Mount Waskey in the Wood River Mountains. A fifth glacier, Chikuminuk, largest in the Wood River Range, was mapped during the International Geophysical Year by the American Geographical Society. At that time (1958), the glacier was three miles long, about a half-mile wide, and had an area of less than two square miles. When mapped, much of the glacier's surface was bare ice with a very small area of accumulation. Trimline positions indicated that Chikuminuk Glacier was receding.

Chikuminuk Glacier, about three miles long, is the largest in the Wood River Mountains. The glacier was mapped in 1958 by the American Geographical Society, whose report observed that much of the glacier's surface was bare ice, and that trimline positions indicated that Chikuminuk was retreating. (Reproduced with permission of the American Geographical Society)

Seward Peninsula

Today only a very small area of the Seward Peninsula supports glaciers. Three locations in the Kigluaik Mountains contain a total of less than two square miles of glacier ice. The largest glacier, Grand Union, in Grand Union Creek drainage, is only about a half-mile long. Two other small cirque glaciers have been identified in Glacier Lake Valley and at Smith Creek.

At the turn of the century the glacier-covered area of the Kigluaik Mountains was more extensive. A.H. Brooks of the U.S. Geological Survey described two distinct areas, both near the head of the North Fork of the Grand Central River, that contained glaciers in 1900. A third small glacier was identified at the head of Pass Creek 13 years later. David M. Hopkins of the U.S. Geological Survey examined aerial photographs taken in 1949 and 1950 of the three areas. Hopkins found that all the glaciers previously described had completely melted away. Three previously unknown glaciers, including Grand Union Glacier, were identified in the aerial photographs.

Hopkins observed Grand Union Glacier in 1973. In the 23 years between his field observations and the 1950 aerial photographs, Grand Union Glacier had decreased in length by 50%.

Left — *A small cirque glacier lies near the Grand Union Creek drainage, north of Mount Osborn (4,714 feet), in the Kigluaik Mountains. Today, the mountains contain a total of less than two square miles of glacier ice.* (Alison Till)
Below — *This photo, taken in 1900 by A.H. Brooks of the U.S. Geological Survey, shows a large cirque glacier at the head of the North Fork of the Grand Central River, in the Kigluaik Mountains. At that time, Brooks found only two areas in the mountains which supported glaciers.* (Brooks, 360, U.S.G.S.)

Brooks Range

The Brooks Range, northernmost mountain range in Alaska, extends for 600 miles from Canada to the Chukchi Sea. The highest peaks in the range rise from the Franklin and Romanzof mountains with Mount Chamberlin (9,020 feet), Mount Isto (9,060 feet), and Mount Hubley (8,915 feet) the highest summits. In addition, three other mountain groups in the range — Schwatka, Endicott, and Philip Smith — support glaciers.

Two separate climatic systems affect the Brooks Range. The northern part of the range is subject to an arctic climate, with average annual temperatures below freezing, while the southern part of the range is subject to the Interior's continental climate. At Anaktuvuk Pass (2,200 feet) the mean annual temperature is 14°F., with precipitation of about 10 inches per year.

Stephen Porter of the University of Washington, who has studied many glaciers in the Anaktuvuk Pass area, reports that many of these ice masses lie in deep, north-facing cirques at elevations below the present regional snowline. He concludes that they result from climatic conditions that do not exist at present.

In 1958 Robert L. Detterman, Arthur L. Bowsher, and J. Thomas Dutro, Jr., of the U.S. Geological Survey determined that six distinct glacial episodes had occurred on the north slope of the Brooks Range. Their sequence is based on analysis of moraines and other glacial deposits. The Anaktuvuk Glaciation, their oldest, dates from the middle Pleistocene and includes moraines that extend 40 miles north and 25 miles south of the mountain fronts.

The Itkillik Glaciation bridges the last part of the Pleistocene into the Holocene and is the best studied of the Brooks Range glacial episodes. Porter and Thomas Hamilton have used radiocarbon dating to subdivide and refine the sequence of events of the Itkillik Glaciation. During this glaciation's maximum extent, the Brooks Range supported a continuous and extensive mountain glacier complex that was almost 500 miles long and 150 miles wide. A separate smaller mountain glacier system, 120 miles by 50 miles, covered the DeLong Mountains to the west.

The youngest glacial event, the Fan Mountain Glaciation, included glacial advances during Neoglacial time. Today, Brooks Range glaciers cover an area of less than 1,000 square miles.

Little knowledge existed about the distribution of Brooks Range glaciers until after World War II. Philip Smith and J.B. Mertie, Jr., reported that as of 1930 only three glaciers had been seen in all of northwestern Alaska. In July 1911 Smith photographed two of these glaciers near the Arrigetch Peaks. In the eastern Brooks Range the only early description of glaciers was in observations made by Ernest Leffingwell in June 1907.

The largest glaciers and the greatest concentration of them within the Brooks Range flow from the Romanzof and Franklin mountains east of the Canning River, within the Arctic National Wildlife Refuge. About 100 square miles of the Romanzofs are glaciated and at least five glaciers have lengths of five miles or more. Two, Okpilak and McCall, were investigated in the late 1950s.

Okpilak Glacier, visited by Leffingwell in 1907, is about five and one-half miles long and covers an area of about five square miles. The ice mass descends from an unnamed, 8,060-foot-high mountain 11 miles south of Mount

Above — *The craggy Arrigetch Peaks, in the Endicott Mountains, support a good number of small cirque and valley glaciers, all unnamed. Most of the glaciers are restricted to elevations above 6,000 feet and show signs of thinning and retreating.* (John and Margaret Ibbotson)
Right — *The summit of Mount Chamberlin (9,020 feet), in the Franklin Mountains, is heavily covered with glaciers, the two largest of which are each less than two miles long.* (David Moyer)

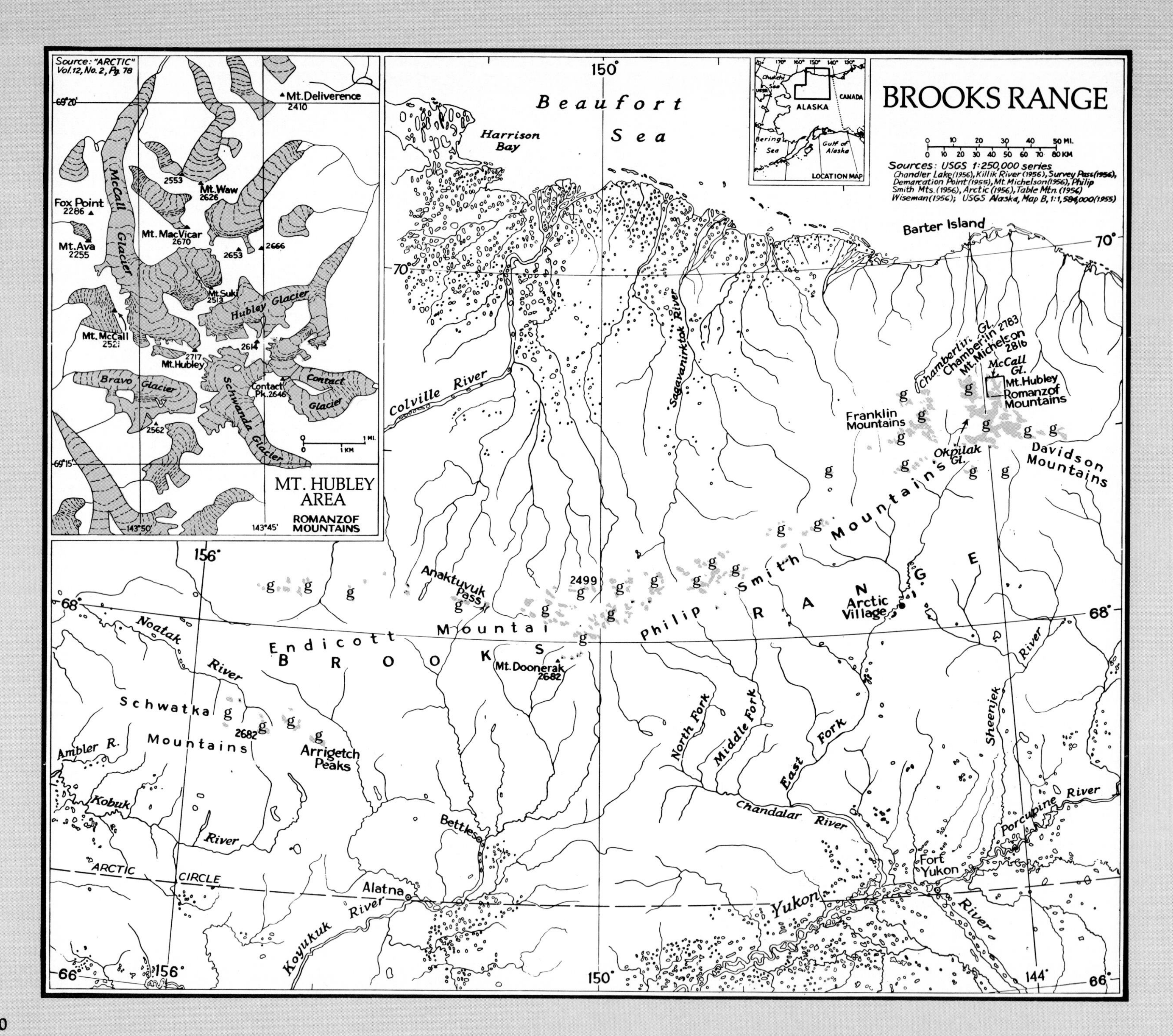
BROOKS RANGE
MT. HUBLEY AREA
ROMANZOF MOUNTAINS
Source: "ARCTIC" Vol.12, No. 2, Pg. 78
Sources: USGS 1:250,000 series
Chandler Lake (1956), Killik River (1956), Survey Pass (1956), Demarcation Point (1955), Mt. Michelson (1956), Philip Smith Mts. (1956), Arctic (1956), Table Mtn. (1956) Wiseman (1956); USGS Alaska, Map B, 1:1,584,000 (1955)
LOCATION MAP
Beaufort Sea
Harrison Bay
Barter Island
Colville River
Sagavanirktok River
Franklin Mountains
Davidson Mountains
Romanzof Mountains
Mt. Hubley
Mt. Michelson 2816
Philip Smith Mountains
Endicott Mountains
BROOKS RANGE
Anaktuvuk Pass
Arctic Village
Mt. Doonerak 2682
Noatak River
Schwatka Mountains
Arrigetch Peaks
Ambler R.
Kobuk River
Bettles
Alatna
Koyukuk River
ARCTIC CIRCLE
Chandalar River
North Fork
Middle Fork
East Fork
Sheenjek River
Porcupine River
Fort Yukon
Yukon River

Michelson (8,855 feet). In 1958 Edward Sable examined a number of Romanzof Mountain glaciers and revisited photographic sites used by Leffingwell. Comparison photographs document that Okpilak Glacier has retreated about 1,000 feet in the intervening 51 years. Sable comments that all glaciers comparable in size to Okpilak show evidence of recent retreat and thinning.

McCall Glacier, 10 miles northeast of Okpilak, heads on Mount Hubley. The glacier, named for Dr. John McCall, former head of the Geology Department at the University of Alaska, has been studied more intensively than any other glacier in the Brooks Range. This research was started by Dr. Richard Hubley during the International Geophysical Year (1957-1958) and was interrupted by his untimely death in an accident on the glacier in October 1957. Between 1969 and 1975 research was continued on heat flow and the ice, water, and mass balance of the glacier by Dr. Gerd Wendler and others at the University of Alaska as part of the International Hydrological Decade. During the four years of observation (1969 through 1972), McCall Glacier's mass balance was consistently negative. Ninety-eight percent of ice loss was due to melting and only two percent to evaporation. These are mass balance conditions typical of arctic glaciers.

Other mountains in the Romanzofs that support large valley glaciers are Mount Isto (9,060 feet), Tugak Peak (8,500 feet), and Mount Michelson (8,855 feet). The summit of Mount Chamberlin (9,020 feet), in the Franklin Mountains, is also heavily glaciated. The two largest glaciers that descend from the summit, Chamberlin Glacier on the west and an unnamed glacier on the east, are both less than two miles in length.

Half-mile-wide Chamberlin Glacier was investigated in 1958 by the U.S. Air Force Cambridge Research Center. Work centered on the analysis of meteorological conditions and a survey of glacial geology.

Glaciers of the Endicott and Philip Smith mountains have been studied by James Ellis and Parker Calkin of the State University of New York at Buffalo and by Thomas Hamilton of the University of Alaska. In 1962 Hamilton reoccupied Philip Smith's 1911 photographic stations and made photographic comparisons of position and health of the glaciers in the Arrigetch Peaks area of the Endicott Mountains. Hamilton noted that a recession and thinning of the glaciers had occurred in the 51 years between studies. Most glaciers in the area are two miles or less in length and are at elevations of 6,000 feet or greater.

In the last few years, Ellis and Calkin have examined an area around the trans-Alaska pipeline near Atigun Pass. They identified 133 glaciers, the largest of which is about one and one-half miles long. All of these glaciers had their termini at elevations above 5,000 feet. As in other parts of the Brooks Range, the glaciers they observed were wasting away. During July and August of 1977, Ellis and Calkin observed the loss of the entire previous year's snow accumulation, plus the loss of an additional three to six feet of ice at three cirque glaciers near Atigun Pass.

In the Schwatka Mountains, south of the Noatak River, at least six peaks support valley glaciers. The largest of these descend from the slopes of 8,510-foot Mount Igikpak and 7,310-foot Oyukak Mountain. The largest ice mass on Igikpak is two miles long, unnamed, and terminates at an elevation of about 3,800 feet.

The largest glacier in the Schwatka Mountains, in the central Brooks Range, is located on Mount Igikpak (8,510 feet). The glacier, which is unnamed, is about two miles long and terminates at an elevation of 3,800 feet.
(John and Margaret Ibbotson)

Where To See Alaska's Glaciers

Editor's note: *Travel to view glaciers is part of enjoying the Alaska experience, but travel* on *a glacier is a different matter which should be undertaken only with thorough preparation and knowledge of the risks involved.*

The number of different glaciers that one can see in Alaska depends on time, cost, and effort. Some glaciers are accessible by public transportation, whereas visits to more remote glaciers may require months of preparation, complicated logistics, favorable weather, and great expense.

Alaska Marine Highway ferries sail past many glaciers along their route through the Inside Passage. Private concessionaires and guides sail to Columbia and other tidewater glaciers in Prince William Sound, and provide tours of Glacier Bay and Frederick Sound. Charter buses make daily trips to Portage Glacier and Mendenhall Glacier.

Charter air services fly over the ice masses of Glacier Bay, Sargent Icefield, Harding Icefield, Stikine Icefield, the Saint Elias Mountains, and the Alaska Range. Cruise ships depart Seattle, Vancouver, and many southeast Alaska cities

Opposite — *Many charter boat services are available to those wishing to observe glaciers. Here, the* Glacier Queen *plies the waters of Columbia Bay, in front of the immense terminus of Columbia Glacier.* (Tom Walker)
Left — *Charter air services provide the opportunity to fly over many of Alaska's glaciers. The charters are also used by experienced climbers to fly into more remote areas.* (Vincent McClelland)

frequently during the summer for Glacier Bay. About 100,000 people a year visit and view the glaciers of Glacier Bay without leaving the decks of these ships.

Several of the state's glaciers are close to frequently traveled roads and can be safely viewed by automobile. Mendenhall Glacier is just 13 miles from downtown Juneau, at the end of Mendenhall Glacier Spur Road. Within driving distance of Anchorage are Matanuska Glacier, located on the Glenn Highway about 100 miles northeast of Anchorage, and Portage Glacier, one of the area's most popular tourist attractions. To reach the glacier, drive southeast on the Anchorage-Seward Highway to Milepost 48.1 and turn off on the Portage Glacier access road. The U.S. Forest Service Visitor Center, located about five miles down the road at the edge of Portage Lake, offers a view of the glacier's terminus and iceberg-filled lake, is a center for nearby nature walks, and has a staff to

Visitors to Portage Glacier inspect a piece of glacial ice with the help of a U.S. Forest Service employee. The forest service operates a visitor center at the glacier, which is staffed with knowledgeable personnel daily from 9 A.M. to 6 P.M. during the summer months. (Third Eye Photography)

provide information and answer questions. Two glaciers are easily visible from the Copper River Highway, which begins in Cordova: Sheridan Glacier can be seen from the bridge over Sheridan River at Milepost 14.8; and Childs Glacier, an active, calving glacier, can be seen at Milepost 38.8. Worthington Glacier, 30 miles east of Valdez, is frequently visited by motorists because of its proximity to the Richardson Highway — the glacier's terminus is less than one-half mile from the highway.

Muldrow Glacier in Denali National Park and Preserve, Valdez Glacier, many Chugach Mountains glaciers east of Anchorage, and those along the Copper River can be visited by foot.

Many glaciers of the Juneau Icefield can be reached by trails maintained by the U.S. Forest Service. At Mendenhall Glacier, three hiking trails provide opportunities to see the component parts of the glacier and evidence of its activity and erosional and depositional history. The Trail of the Glacier is a one-half-mile, self-guided nature trail that has 15 numbered stops, each demonstrating a different facet of the activities of the glacier, or providing some useful information about fauna and flora that now occupy Mendenhall Valley. A handy, free guide to the Trail of the Glacier, available from the visitor center, makes this one of the most intriguing and informative ways to observe the effects of recent glaciation.

East Glacier and West Glacier trails, three and six miles long respectively, permit hikers to walk over recently deglaciated terrain to positions overlooking the active moving glacier. Access to the glacier is possible from West Glacier Trail, but is not recommended unless the hiker possesses skill and equipment for glacier travel. East Glacier Trail provides excellent views of the eastern part of Mendenhall's terminus, and of Nugget Creek, which runs along the southeast side of the glacier. There is no other location in Alaska that possesses such a complete variety of glacial components, activities, and products as does

A young boy enjoys the tranquility of his surroundings at Worthington Glacier, located just one-half mile off the Richardson Highway. Because it is so easily accessible, Worthington is often visited by motorists traveling to Valdez. (Helen Rhode)

Figure 9 — *Trails in the Mendenhall Glacier complex in southeastern Alaska.*

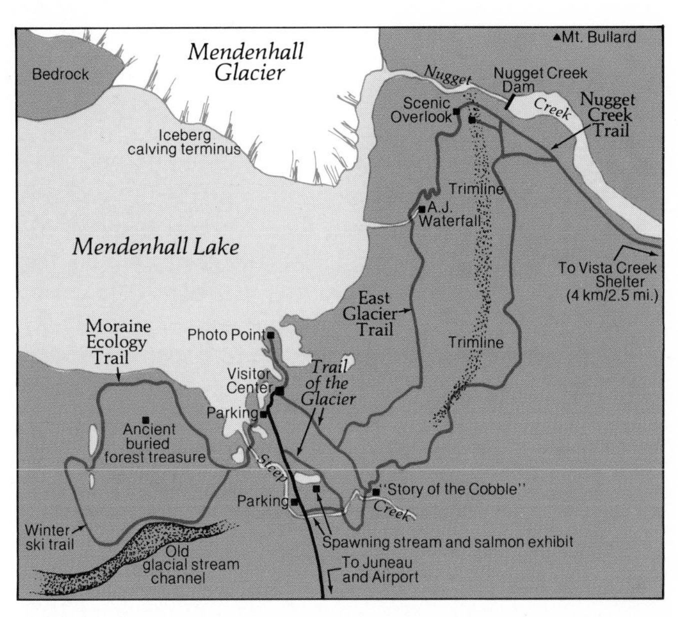

Mendenhall Glacier with its visitor center, trails, and remarkably easy access.

Complete information on how to get to Alaska's more accessible glaciers, as well as names of charter bus and boat services, and railroad service, can be found in *The MILEPOST®*, available from Alaska Northwest Publishing Company, P.O. Box 4-EEE, Anchorage, AK 99509. Write for current price.

Information on travel in Alaska can also be obtained through the State of Alaska Division of Tourism, Pouch 'E', Juneau, AK 99811. Ferry schedules are available from the Marine Highway System, Pouch 'R', Juneau, AK 99811.

Glacier Travel

Glacier travel is not for the uninitiated. It is dangerous, sometimes fatal, and is one of the primary activities that leads to hypothermia. Do not travel alone on a glacier; if you are inexperienced, do not travel there at all. Glacier accidents take their toll on the experienced traveler as well as the novice. Almost every year the lives of experienced climbers and hikers are lost to hypothermia, in avalanches, or in crevasse or icefall accidents. The experienced glacier traveler knows the risks involved in glacier travel; beginners do not.

Left — *A group of climbers waves to the pilot who flew them in to Gerstle Glacier, in the Alaska Range. When exploring a glacier, be sure to always carry emergency shelter, dry clothing, and extra food.* (Warren Yeend, U.S.G.S.)
Above — *Two climbers stand high above a gaping crevasse in Glacier Bay National Park and Preserve. Extreme caution should be exercised around crevasses — every year injury and even death result from accidents involving crevasses.* (Tom Bean)

Left — *Two rules for glacier travelers are to never travel alone, and use skis or snowshoes whenever possible. Here, Pete Robinson checks his map while companion Bev Lawrence approaches.* (Mark Skok)
Below — *The primary source of danger to kayakers traveling near the face of a glacier is falling ice. Here, a large chunk of Margerie Glacier, in Glacier Bay National Park and Preserve, crashes into Tarr Inlet.* (Dennis Hellawell)

If you plan to hike in a glacier environment, keep the following rules in mind:

1. Do not travel alone in a glacier environment.
2. Always travel with an experienced companion.
3. Know the symptoms of hypothermia and know how to treat it.
4. In crevasse areas, always rope-up and always probe for snow bridges.
5. Avoid icefalls, *seracs,* and avalanche areas.
6. Always carry emergency shelter, dry clothing, and extra food.
7. Know the fundamentals of using crampons, ice ax, and rope.
8. Whenever possible, travel on skis or snowshoes as they distribute body weight over a much greater area than travel by foot.
9. Do not travel in whiteout conditions. Make camp and wait for the weather to clear.
10. Use common sense and do not panic.

Cheechako At Childs Glacier

By Janet O. Cogan

With my husband Jim as assistant, I temporarily replaced the regular veterinarian in Cordova and Valdez on Prince William Sound in summer 1979.

In Cordova, many local people live by an unwritten rule which provides for an automatic day off every time the sun shines. After three weeks of uninterrupted rain and work, we set forth under clear skies to take advantage of the sunshine law. We packed a picnic lunch and, joined by our friends Sue and Pam, piled into the musty old pickup. The dogs, Bean and Mandy, rode in the bed of the pickup as we headed out the Copper River Highway to Million Dollar Bridge and Childs Glacier.

When we reached the bridge, we turned left onto a one-lane road cut through dense vegetation. We emerged from the willows and poplars at the edge of a steep cut bank. Below us surged the Copper River, icebergs bobbing in its swift current. The imposing fissured snout of Childs Glacier rose above the Copper, its blue ice dominating the scene for two miles downriver.

While Sue was marveling at how much the bank had eroded since her visit the year before, we eagerly spilled out of the pickup cab and climbed down the loose gravel to the water's edge. We had come to watch the glacier. Selecting polished boulders as our grandstand seats, we settled ourselves in the warm July sun.

Within moments a sharp crack, like the report of a rifle, split the air. A heavy creaking and thunderous moan followed as a large slab of ice toppled from its perch and splashed into the river. The river swelled in response to the icefall, and a wave advanced toward us, raising the shoreline about a foot.

Wet and frightened, Bean had somehow made it to the relative safety of the pickup when the wave hit. (Janet Cogan)

The cracking and popping of Childs Glacier continued and irregular blocks of ice fell, joining the flotilla of bergs being carried to the ocean. Sue and Pam told stories of how falling ice had killed people that ventured too close to other glaciers. We all agreed that only fools would tempt the mighty power of a moving river of ice. We felt secure, separated from the glacier by a quarter-mile of swiftly flowing river. We shouted at the wall of ice, hoping the sound vibrations would help dislodge loosened chunks. We got our wish, and then some.

On the icy face opposite us, a blue crack opened and slowly snaked its way up and across the perpendicular surface. Loud cracking soon gave way to a deep rumble as the ice began to calve. The entire glacier wall was crumbling and beginning to slide into the river. Ice, peeling away in a shower of sparkling crystals, crashed into the foamy water below and sent a huge wave heaving in our direction. The roar was deafening but no communication was necessary. Sensing the threat to our lives, we quickly rose from our rock perches and moved toward the cut bank behind us. Our scramble up the loose gravel came to a struggling halt just below the edge. My arms reached the top of the bank but I could only claw uselessly at the crumbling, sandy soil. The tangle of poplars, bent flat and completely uprooted, provided no firm handhold. My efforts were further hindered by my awkwardness, being seven-months pregnant.

The others fared no better. Clustered together below the edge of the bank and knowing there was no time for escape, we all turned to watch the advancing wave. Its foam-flecked crest swelled before our eyes.

The wave rose and curled forming a 12-foot wall of water. For a brief moment the wave hung there, poised near the shore. Finally the wave fell, thundering and foaming onto the beach. Boulders churned and clicked as if they were pebbles. The frothy water rushed up the bank to within inches of our feet. Abruptly, the wave vanished, disappearing into the wet, black sand. The boulders we had been sitting on moments before were gone. So was Sue's dog, Bean. We stared in silence, unmoving.

Shock on Sue's face soon gave way to tears as she ran down the beach searching for the black-and-white form of Bean among the rubble on shore. Jim joined her in the search. Pam and I found a spot where we could scale the bank and, with our dog Mandy, we struggled through the dense, tangled trees toward the truck. When we reached it, we discovered the soaked, bruised, and shaking Bean, curled on the floor of the pickup cab. Somehow the dog had escaped the wave. Our holler of surprise and joy signaled Sue and Jim and they came running. We were all eager to leave.

I felt light-headed, almost giddy, after our brush with death. My sharpened senses were now acutely aware of the signals nature had given us which we had foolishly chosen to ignore. The freshly eroded bank, the uprooted trees, the loose rounded boulders, all were clues to the power behind the icy muscles of the glacier. Looking at these scars around us, it was obvious that many waves washed completely over the bank.

Suddenly it happened: a piece of ice the size of a football field sheared away with a roaring crack and began falling, as if in slow motion. Everyone froze in amazement momentarily, then began scattering for any safe place they could reach, leaving their belongings where they lay. I pulled my camera to my eye and began recording the event as fast as I could advance and release the shutter. The wave that rolled toward the shore seemed small in the viewfinder, so I continued snapping pictures. After a few shots, I heard the splash of a wave and looked up to see it break on the shore about 50 feet from me and coming my way. I remained where I stood, but planned my escape to the safety of the far side of a van parked about 15 feet behind me. The wave continued to advance, and then, with the sound of an ocean wave pounding the beach, it hit the embankment in front of me and splashed up in a spray of muddy water, rocks and ice. I dove for my shelter just as the water splashed up over the top of the van. Drenched by the cold water and peppered by the particles it carried, I found myself running for the bushes behind me as the water swirled around my shins. Soon the flood crested and the water began slowly to recede. As I headed back to the truck to see how the others had fared, I saw Bert, one of my companions, standing with arms outstretched, looking at his thoroughly soaked clothes with an astonished face. The wave had knocked him down, but he was able to grab the truck's door handle as he washed past, causing the water to rush over him. (Photos and caption by Ron Rose)

Glossary

ABLATION — The combination of processes by which a glacier loses ice and snow. Melting, sublimation (evaporation), calving, and wind transport (erosion) are the processes that remove water from a glacier system.

ALTITHERMAL — The period of time during the Holocene from about 6,600 years ago to about 5,300 years ago. During this interval, climate was drier and warmer than the present, with temperatures as much as three to five degrees Farenheit warmer. Most of Alaska's glaciers retreated during this period.

ARÊTE — The jagged, narrow ridge that separates two adjacent glacier valleys or cirques. The ridge, which resembles the blade of a serrated knife, is termed *arête* after the French word for fish bones. Photo shows *arête* ridge surrounding Amherst Glacier in College Fiord.

BERGSCHRUND — A single, large crevasse or series of smaller crevasses that develop at the head of a valley glacier at the point where moving ice pulls away from stagnant ice or the rock of the cirque wall against which it accumulated. In winter, the *bergschrund* is filled with snow, but in summer it reopens and is visible. Example is in the Juneau Icefield.

BERGY SELTZER — A crackling or sizzling sound, similar to champagne, seltzer water, or Rice Krispies, that is produced by an iceberg while it is melting. The sound is caused by the release of air bubbles that were trapped in glacial ice during its formation under extremely high pressure. Synonymous with ICE SIZZLE.

Steve McCutcheon

CALVING — The process by which masses of ice break away from the terminus of a glacier or ice sheet that ends in a body of water. Icebergs produced in this manner drift away under the influence of wind, waves, tides, and currents. This is Hubbard Glacier.

Tom Smayda

CIRQUE — A bowl-shaped, amphitheater-like depression eroded into the head or high on the side of a glacier valley. Photo is north fork of Ruth Glacier.

CREVASSE — Narrow, deep, elongate cracks that develop in the surface of a moving glacier in response to stresses created by movement over uneven or irregular subsurfaces. Many are concealed by covers of snow called snow bridges. Photo shows the surface of a glacier in the Juneau Icefield.

DISTRIBUTARY — A stream or glacier that flows away from the main stream or the main glacier. In certain glaciers in Alaska, ice streams that at one time supplied ice to a larger glacier as a tributary now, because of accelerated melting and downwasting, are at lower elevations than the main glacier. In these instances, ice flows from the main glacier into these shrinking distributary systems. In streams, any channel that leaves the main stream and does not reenter it is a distributary.

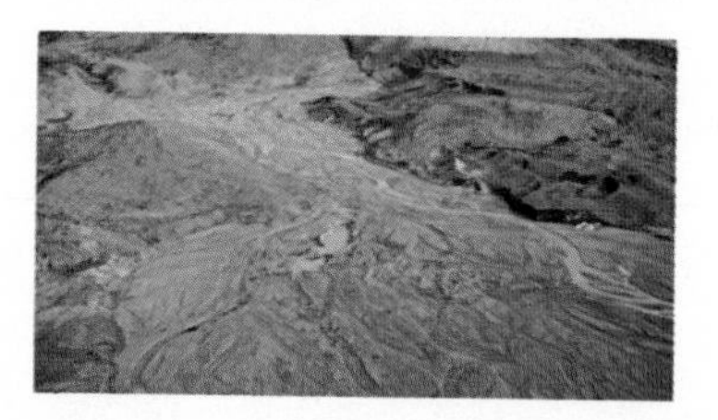

DOWNWASTING — The process of a glacier thinning during ablation. The term often describes the melting of a stagnant block of ice or stagnant glacier. Example shows Casement Glacier.

DRIFT — A collective term used to describe all sedimentary deposits of a glacier, regardless of the size of the material involved. The term includes all material that is transported by a glacier, deposited directly by a glacier, or deposited by running water that originates from a glacier.

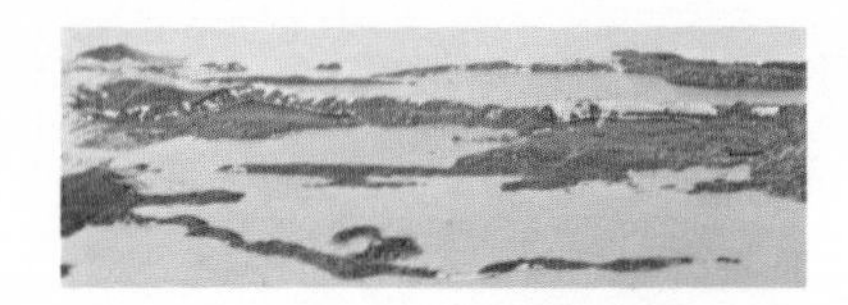

ESKER — A meandering, steep-sided sediment ridge that forms either beneath or within a glacier by the depositional action of meltwater streams. Eskers are generally composed of sand and gravel. They range from feet to miles in length, and may exceed 100 feet in height. These eskers were left by melting ice in Malaspina Lake.

EUSTACY — A term used to describe worldwide sea level conditions, and the worldwide sea level regime. When continental ice sheets formed, large volumes of water were removed from the world's oceans to form glaciers. This worldwide lowering of sea level is known as "a eustatic sea level lowering."

FIORD — A glacially eroded or modified U-shaped valley that extends below sea level into a deep bay or onto the continental shelf. Depths frequently exceed 1,000 feet below sea level. Widths may be up to five miles with lengths of more than 100 miles not uncommon. A fiord may have a shallow lip at its seaward end. All fiords are the result of glacial erosion. Photo shows Muir Inlet.

Steve McCutcheon

FIRN — An intermediate stage in the transformation of snow to glacier ice. Snow becomes firn when it has undergone compression so that no pore space remains, a process that takes about one summer. Snow has a density of 0.1 to 0.3 grams per cubic centimeter, while firn's density is greater than 0.4. Photo is Casement Glacier in Glacier Bay.

FOLIA — Individual layers or bands of ice crystals produced during the metamorphic process that changes snow to glacier ice.

FOLIATION — The layering or banding that forms within a glacier as it undergoes the metamorphism that changes snow to glacier ice. Layers generally alternate between bubbly and clear ice or between fine-grained and coarse-grained ice. Example shows the terminus of Lamplugh Glacier.

GLACIER — A large, perennial accumulation of ice, snow, rock, sediment, and water that originates on land and moves down slope under the influence of its own weight and gravity.

Glacier remanié — A glacier formed by the reconstitution through pressure melting (regelation) of ice blocks that accumulate beneath or below the ends of glaciers that are at higher elevations. Avalanches and icefalls bring the ice blocks down from higher elevations.

Betty Johannsen

Hanging glacier (example at right) — A glacier that originates high on the wall of a U-shaped valley and descends only part way to the level of the main glacier. Ice avalanching is the primary method that transports material from the hanging glacier to the valley below. This is Explorer Glacier.

HANGING VALLEY — A tributary glacial valley that is perched high on the wall of a larger U-shaped glacial valley. The hanging valley floor may be hundreds of feet above the floor of the main valley. When both valleys were filled with ice, the glacier that filled the hanging valley probably flowed into the upper part of the glacier that filled the large valley. With the melting of both glaciers, the bedrock valleys they occupied are no longer connected. Streams in the hanging valley enter the main valley as waterfalls.

HOLOCENE — The period of time following the end of Pleistocene. This interval, which includes the present, began between 15,000 and 10,000 years ago.

Gil Mull

HORN — A steep-sided, sharp-pointed pyramidal mountain peak formed by three or more glaciers expanding their valleys and cirques headward or upglacier. Example is Mount Chamberlin, 9,000 feet in Franklin Mountains.

ICE SIZZLE — See BERGY SELTZER.

ISOSTACY — A process whereby a balance of mass is maintained within the earth's crust and mantle in response to an increase or decrease of mass at a point on the earth's surface. When glaciers sat on the surface of an area the elevation of that area was depressed. When the glaciers retreated, the area rose in response to the decrease in surface mass.

Gil Mull

JOKULHLAUP — An Icelandic term for glacier outburst floods. Glacial floods in Alaska result from the failure of ice dams more frequently than from any other cause. This *jokulhlaup* occurred when Hidden Lake emptied into the Kennicott River.

KAME — A poorly sorted sand and gravel deposit which forms where running meltwater encounters stagnant ice. Kames generally form within cracks, holes, or crevasses in the ice, or between the ice and the adjacent land surface. Kames may be hills, mounds, knobs, hummocks, or ridges.

KAME TERRACE — A terrace that consists of stratified sand and gravel that forms between the glacier or a stagnant ice block and the valley wall.

KETTLE — A depression that forms on an outwash plain or in other glacial drift by the melting of a block of ice that was left behind during a glacier's retreat, or by a block of ice that was floated to the outwash plain by a meltwater stream and then buried by later sedimentation.

LITTLE ICE AGE — The period beginning about 5,500 years ago and ending in some areas as recently as 1550 to 1850 A.D. during which mountain glaciers in many parts of the world, including Alaska, experienced significant expansions.

MATTERHORN — A glacial horn formed by the symmetrical erosion of four separate glaciers. Named after the famous peak in the Pennine Alps.

MORAINE — An unstratified and unsorted glacial deposit that forms through direct action or contact with glacier ice. Example above shows the lower portion of Topeka Glacier's valley.

Ablation moraine — A layer or pile of ablation till formed by the melting of a stagnant ice mass. Shown is Lucia Glacier in the St. Elias Mountains.

Ground moraine — A blanket of till deposited on all surfaces over which a glacier flows.

Lateral moraine (example above) — An accumulation of glacial material at the side margins of a mountain glacier. This is Muir Glacier.

Tom Bean

Medial moraine — A moraine that occupies a central region of a mountain glacier, formed by the coalescing of two lateral moraines when two valley glaciers merge. This is McBride Glacier in Glacier Bay.

Push moraine — A pile of glacial material that is bulldozed forward by the advancing terminus of a glacier. This push moraine is located in front of Reid Glacier in Glacier Bay.

Recessional moraine — An accumulation of glacial material that forms while the terminus of a retreating glacier remains at or near a fixed position. A series of such moraines may represent a number of standstills during a glacier's retreat.

Terminal moraine — The farthest downvalley moraine that accumulates during a specific period of a glacier's advance. This recessional moraine is also known as an end moraine. Example is Reid Glacier.

MOULIN — Narrow tubular chutes in a glacier through which streams flow. Also called glacier mills. This *moulin* drains water from La Perouse Glacier.

NEOGLACIATION — The re-advance of mountain glaciers during the Little Ice Age.

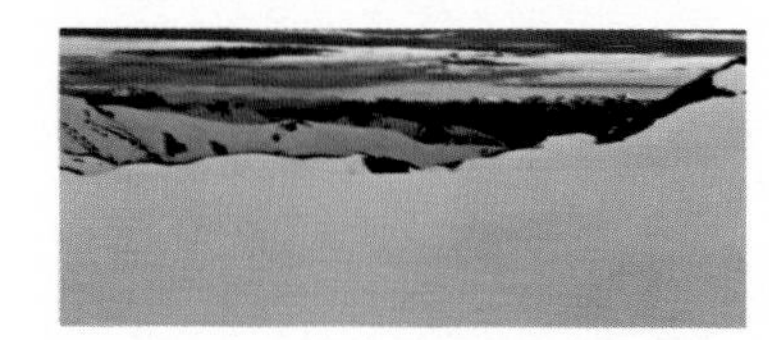

NÉVÉ — The area of a glacier covered with perennial snow. The term is also used to describe the area of accumulation at the upper end of a glacier. Example above is in the Juneau Icefield.

NUNATAK — Mountain peaks and ridges which poke through the ice, separating large interconnecting valley glaciers. Photo shows Riggs Glacier.

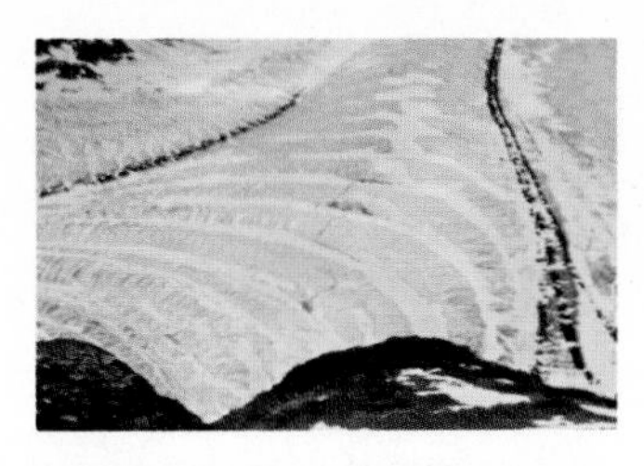

OGIVE — Curved bands or undulations which form in the ice at the surface of a glacier, usually at the base of an icefall. These bands are generally oriented in a convex, downglacier position. Ogives occur in two types: band ogives, which are alternating light and dark bands on a flat, smooth glacier surface; and wave ogives, which are undulations of varying height at the ice surface. Example shows ogives on Vaughn Lewis Glacier.

OUTWASH PLAIN — A broad, gentle, alluvial plain composed of glacially eroded material transported by meltwater. The plain begins at the edge of a retreating glacier and may extend for many miles. Material in the plain is stratified and somewhat sorted. This is Harriman Glacier northeast of Whittier.

PLEISTOCENE — The epoch of geological time that began more than two million years ago and ended about 10,000 to 15,000 years ago. The Pleistocene is popularly known as "The Great Ice Age" or the "glacial epoch." Together, the Pleistocene and the Holocene comprise the Quaternary Period.

PLUCKING — The removal and transportation by a moving glacier of large blocks of rock. These blocks are "quarried" and prepared for removal by the freezing and thawing of water in cracks, joints, and fractures.

RECONSTITUTION — The process that recombines ice blocks that have avalanched from higher elevations into a new glacier. See *Glacier remanié.*

ROCHE MOUTONNÉE — Glaciers sculpt the bedrock surfaces over which they move. One of the most common forms produced is a large, rounded, asymmetrical bedrock knob with a gentle slope on the upglacier side and a steep to almost vertical face on the downglacier side. These bedrock knobs are *roches moutonnées.* Photo shows Marble Island in Glacier Bay.

ROCK FLOUR — Finely ground, silty rock particles produced by abrasion at the base and sides of a glacier. Rock flour is responsible for coloring streams that drain glaciers either a milky white, gray, or brown. Example shows rock flour at Lamplugh Glacier.

SÉRAC — A jagged pinnacle or tower of glacier ice that comprises the surface of a glacier. Some *séracs* form where crevasses intersect. Photo is McBride Glacier.

SURGE — Sudden large-scale, short-lived movements within a glacier are termed surges. Surges are periodic, do not necessarily involve an advance of the glacier terminus, and consist of movements that are 10 to more than 100 times faster than a glacier's normal rate of flow.

TERMINUS — The lower margin or extremity of a glacier. Also referred to as the snout or the toe. Example shows McBride Glacier.

Lobate terminus — A terminus consisting of one or more rounded lobes. This is Taku Glacier.

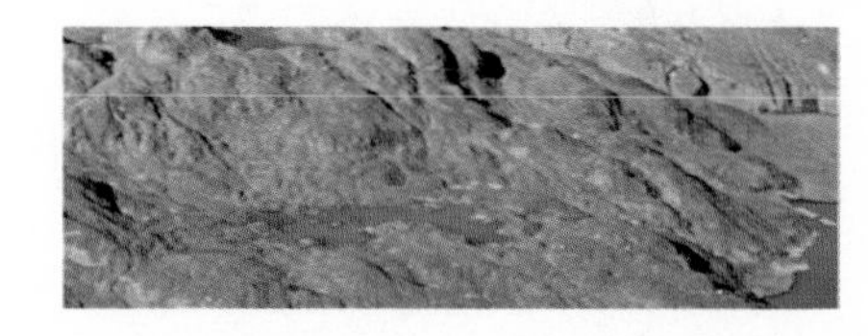

TILL — Any unsorted and unstratified glacial drift that was deposited directly by a glacier and was not reworked by water. Till contains a great variety of different-sized material in a heterogeneous mixture. Example is at Muir Inlet, Glacier Bay.

Lodgement till — An unsorted glacial deposit plastered in place by a moving glacier. The deposit is unsorted but may show some evidence of an orientation.

TRIMLINE — A very sharp boundary line that marks the maximum lateral extent of a glacier. The line may be a color change in the bedrock showing the boundary between weathered and unweathered rock, or may be the change from vegetation to no vegetation. This is Meade Glacier near Skagway.

Wesley Wallace

Alaska Geographic® Back Issues

The North Slope, Vol. 1, No. 1. Charter issue of *ALASKA GEOGRAPHIC®*. Out of print.

One Man's Wilderness, Vol. 1, No. 2. The story of a dream shared by many, fulfilled by few: a man goes into the bush, builds a cabin and shares his incredible wilderness experience. Color photos. 116 pages, $9.95.

Admiralty . . . Island in Contention, Vol. 1, No. 3. An intimate and multifaceted view of Admiralty: its geological and historical past, its present-day geography, wildlife and sparse human population. Color photos. 78 pages, $5.00.

Fisheries of the North Pacific: History, Species, Gear & Processes, Vol. 1, No. 4. Out of print.

The Alaska-Yukon Wild Flowers Guide, Vol. 2, No. 1. First Northland flower book with both large, color photos and detailed drawings of every species described. Features 160 species, common and scientific names and growing height. 112 pages, $10.95

Richard Harrington's Yukon, Vol. 2, No. 2. Out of print.

Prince William Sound, Vol. 2, No. 3. Out of print.

Yakutat: The Turbulent Crescent, Vol. 2, No. 4. Out of print.

Glacier Bay: Old Ice, New Land, Vol. 3, No. 1. The expansive wilderness of Southeastern Alaska's Glacier Bay National Monument unfolds in crisp text and color photographs. Records the flora and fauna of the area, its natural history, with hike and cruise information, plus a large-scale color map. 132 pages, $9.95.

The Land: Eye of the Storm, Vol. 3, No. 2. Out of print.

Richard Harrington's Antarctic, Vol. 3, No. 3. The Canadian photojournalist guides readers through remote and little understood regions of the Antarctic and Subantarctic. More than 200 color photos and a large fold-out map. 104 pages, $8.95.

The Silver Years of the Alaska Canned Salmon Industry: An Album of Historical Photos, Vol. 3, No. 4. Out of print.

Alaska's Volcanoes: Northern Link in the Ring of Fire, Vol. 4, No. 1. Scientific overview supplemented with eyewitness accounts of Alaska's historic volcano eruptions. Includes color and black-and-white photos and a schematic description of the effects of plate movement upon volcanic activity. 88 pages, $7.95.

The Brooks Range: Environmental Watershed, Vol. 4, No. 2. Out of print.

Kodiak: Island of Change, Vol. 4, No. 3. Out of print.

Wilderness Proposals: Which Way for Alaska's Lands?, Vol. 4, No. 4. Out of print.

Cook Inlet Country, Vol. 5, No. 1. A visual tour of the region — its communities, big and small, and its countryside. Begins at the southern tip of the Kenai Peninsula, circles Turnagain Arm and Knik Arm for a close-up view of Anchorage, and visits the Matanuska and Susitna valleys and the wild, west side of the inlet. 230 color photos, separate map. 144 pages, $9.95.

Southeast: Alaska's Panhandle, Vol. 5, No. 2. Explores Southeastern Alaska's maze of fjords and islands, mossy forests and glacier-draped mountains — from Dixon Entrance to Icy Bay, including all of the state's fabled Inside Passage. Along the way are profiles of every town, together with a look at the region's history, economy, people, attractions and future. Includes large fold-out map and seven area maps. 192 pages, $12.95.

Bristol Bay Basin, Vol. 5, No. 3. Explores the land and the people of the region known to many as the commercial salmon-fishing capital of Alaska. Illustrated with contemporary color and historic black-and-white photos. Includes a large fold-out map of the region. 96 pages, $9.95.

Alaska Whales and Whaling, Vol. 5, No. 4. The wonders of whales in Alaska — their life cycles, travels and travails — are examined, with an authoritative history of commercial and subsistence whaling in the North. Includes a fold-out poster of 14 major whale species in Alaska in perspective, color photos and illustrations, with historical photos and line drawings. 144 pages, $12.95.

Yukon-Kuskokwim Delta, Vol. 6, No. 1. Out of print.

Aurora Borealis: The Amazing Northern Lights, Vol. 6, No. 2. The northern lights — in ancient times seen as a dreadful forecast of doom, in modern days an inspiration to countless poets. Here one of the world's leading experts — Dr. S.-I. Akasofu of the University of Alaska — explains in an easily understood manner, aided by many diagrams and spectacular color and black-and-white photos, what causes the aurora, how it works, how and why scientists are studying it today and its implications for our future. 96 pages, $7.95.

Alaska's Native People, Vol. 6, No. 3. In the largest edition to date — result of several years of research — the editors examine the varied worlds of the Inupiat Eskimo, Yup'ik Eskimo, Athabascan, Aleut, Tlingit, Haida and Tsimshian. Most photos are by Lael Morgan, *ALASKA®* magazine's roving editor, who since 1974 has been gathering impressions and images from virtually every Native village in Alaska. Included are sensitive, informative articles by Native writers, plus a large, four-color map detailing the Native villages and defining the language areas. 304 pages, $19.95.

The Stikine, Vol. 6, No 4. River route to three Canadian gold strikes in the 1800s, the Stikine is the largest and most navigable of several rivers that flow from northwestern Canada through Southeastern Alaska on their way to the sea. This edition explores 400 miles of Stikine wilderness, recounts the river's paddlewheel past and looks into the future, wondering if the Stikine will survive as one of the North's great free-flowing rivers. Illustrated with contemporary color photos and historic black-and-white; includes a large fold-out map. 96 pages, $9.95.

Alaska's Great Interior, Vol. 7, No. 1. Alaska's rich Interior country, west from the Alaska-Yukon Territory border and including the huge drainage between the Alaska Range and the Brooks Range, is covered thoroughly. Included are the region's people, communities, history, economy, wilderness areas and wildlife. Illustrated with contemporary color and black-and-white photos. Includes a large fold-out map. 128 pages, $9.95.

A Photographic Geography of Alaska, Vol. 7, No. 2. An overview of the entire state — a visual tour through the six regions of Alaska: Southeast, Southcentral/Gulf Coast, Alaska Peninsula and Aleutians, Bering Sea Coast, Arctic and Interior. Plus a handy appendix of valuable information — "Facts About Alaska." Approximately 160 color and black-and-white photos and 35 maps. 192 pages, $14.95.

The Aleutians, Vol. 7, No. 3. The fog-shrouded Aleutians are many things — home of the Aleut, a tremendous wildlife spectacle, a major World War II battleground and now the heart of a thriving new commercial fishing industry. Roving editor Lael Morgan contributes most of the text; also included are contemporary color and black-and-white photographs, and a large fold-out map. 224 pages, $14.95.

Klondike Lost: A Decade of Photographs by Kinsey & Kinsey, Vol. 7, No. 4. An album of rare photographs and all-new text about the lost Klondike boom town of Grand Forks, second in size only to Dawson during the gold rush. Introduction by noted historian Pierre Berton: 138 pages, area maps and more than 100 historical photos, most never before published. $12.95.

Wrangell-Saint Elias, Vol. 8, No. 1. Mountains, including the continent's second- and fourth-highest peaks, dominate this international wilderness that sweeps from the Wrangell Mountains in Alaska to the southern Saint Elias range in Canada. The region draws backpackers, mountain climbers, and miners, and is home for a few hardy, year-round inhabitants. Illustrated with contemporary color and historical black-and-white photographs. Includes a large fold-out map. $9.95.

Alaska Mammals, Vol. 8, No. 2. From tiny ground squirrels to the powerful polar bear, and from the tundra hare to the magnificent whales inhabiting Alaska's waters, this volume includes 80 species of mammals found in Alaska. Included are beautiful color photographs and personal accounts of wildlife encounters. *The* book on Alaska's mammals — from Southeast to the Arctic, and beyond! $12.95.

The Kotzebue Basin, Vol. 8, No. 3. Examines northwestern Alaska's thriving trading area of Kotzebue Sound and the Kobuk and Noatak river basins, lifelines of the region's Inupiat Eskimos, early explorers, and present-day, hardy residents. Contemporary color and historical black-and-white photographs illustrate varied cultures and numerous physical attractions of the area. $12.95.

Alaska National Interest Lands, Vol. 8, No. 4. Following passage of the bill formalizing Alaska's national interest land selections (d-2 lands), longtime Alaskans Celia Hunter and Ginny Wood review each selection, outlining location, size, access, and briefly describing the region's special attractions. Illustrated with contemporary color photographs depicting as no other medium can the grandeur of Alaska's national interest lands. $14.95.

COMING ATTRACTION

SITKA, Baranof and Chichagof Islands, Vol. 9, No. 2. From the elegant capital of Russian America to a beautiful but modern port, Sitka, on Baranof Island, has become a commercial and cultural center for southeastern Alaska. Pat Roppel, longtime Southeast resident and expert on the region's history, examines in detail the past and present of Sitka, Baranof Island, and neighboring Chichagof Island. Illustrated with contemporary color and historical black-and-white photographs. To members in May 1982. Price to be announced.

Your $30.00 membership in The Alaska Geographic Society includes 4 subsequent issues of *ALASKA GEOGRAPHIC®*, the Society's official quarterly. Please add $4 for non-U.S. membership.

Additional membership information available upon request. Single copies of the *ALASKA GEOGRAPHIC®* back issues available, per listing here. When ordering please add $1 postage/handling per copy. To order back issues send your check or money order and volumes desired to:

Box 4-EEE, Anchorage, AK 99509